FREE Test Taking Tips DVD Offer

To help us better serve you, we have developed a Test Taking Tips DVD that we would like to give you for FREE. **This DVD covers world-class test taking tips that you can use to be even more successful when you are taking your test.**

All that we ask is that you email us your feedback about your study guide. Please let us know what you thought about it – whether that is good, bad or indifferent.

To get your **FREE Test Taking Tips DVD**, email freedvd@studyguideteam.com with "FREE DVD" in the subject line and the following information in the body of the email:

    a. The title of your study guide.

    b. Your product rating on a scale of 1-5, with 5 being the highest rating.

    c. Your feedback about the study guide. What did you think of it?

    d. Your full name and shipping address to send your free DVD.

If you have any questions or concerns, please don't hesitate to contact us at freedvd@studyguideteam.com.

Thanks again!

DEER PARK PUBLIC LIBRARY
44 LAKE AVENUE
DEER PARK, NY 11729

ACCUPLACER Study Guide 2020-2021

ACCUPLACER Test Prep with Practice Test Questions for All Sections Including Math, English, and Reading [5th Edition]

TPB Publishing

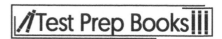

Written and edited by TPB Publishing.

TPB Publishing is not associated with or endorsed by any official testing organization. TPB Publishing is a publisher of unofficial educational products. All test and organization names are trademarks of their respective owners. Content in this book is included for utilitarian purposes only and does not constitute an endorsement by TPB Publishing of any particular point of view.

Interested in buying more than 10 copies of our product? Contact us about bulk discounts:
bulkorders@studyguideteam.com

ISBN 13: 9781628459340
ISBN 10: 1628459344

Table of Contents

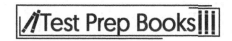

Test Prep Books

Quick Overview

As you draw closer to taking your exam, effective preparation becomes more and more important. Thankfully, you have this study guide to help you get ready. Use this guide to help keep your studying on track and refer to it often.

This study guide contains several key sections that will help you be successful on your exam. The guide contains tips for what you should do the night before and the day of the test. Also included are test-taking tips. Knowing the right information is not always enough. Many well-prepared test takers struggle with exams. These tips will help equip you to accurately read, assess, and answer test questions.

A large part of the guide is devoted to showing you what content to expect on the exam and to helping you better understand that content. In this guide are practice test questions so that you can see how well you have grasped the content. Then, answer explanations are provided so that you can understand why you missed certain questions.

Don't try to cram the night before you take your exam. This is not a wise strategy for a few reasons. First, your retention of the information will be low. Your time would be better used by reviewing information you already know rather than trying to learn a lot of new information. Second, you will likely become stressed as you try to gain a large amount of knowledge in a short amount of time. Third, you will be depriving yourself of sleep. So be sure to go to bed at a reasonable time the night before. Being well-rested helps you focus and remain calm.

Be sure to eat a substantial breakfast the morning of the exam. If you are taking the exam in the afternoon, be sure to have a good lunch as well. Being hungry is distracting and can make it difficult to focus. You have hopefully spent lots of time preparing for the exam. Don't let an empty stomach get in the way of success!

When travelling to the testing center, leave earlier than needed. That way, you have a buffer in case you experience any delays. This will help you remain calm and will keep you from missing your appointment time at the testing center.

Be sure to pace yourself during the exam. Don't try to rush through the exam. There is no need to risk performing poorly on the exam just so you can leave the testing center early. Allow yourself to use all of the allotted time if needed.

Remain positive while taking the exam even if you feel like you are performing poorly. Thinking about the content you should have mastered will not help you perform better on the exam.

Once the exam is complete, take some time to relax. Even if you feel that you need to take the exam again, you will be well served by some down time before you begin studying again. It's often easier to convince yourself to study if you know that it will come with a reward!

Test-Taking Strategies

1. Predicting the Answer

When you feel confident in your preparation for a multiple-choice test, try predicting the answer before reading the answer choices. This is especially useful on questions that test objective factual knowledge. By predicting the answer before reading the available choices, you eliminate the possibility that you will be distracted or led astray by an incorrect answer choice. You will feel more confident in your selection if you read the question, predict the answer, and then find your prediction among the answer choices. After using this strategy, be sure to still read all of the answer choices carefully and completely. If you feel unprepared, you should not attempt to predict the answers. This would be a waste of time and an opportunity for your mind to wander in the wrong direction.

2. Reading the Whole Question

Too often, test takers scan a multiple-choice question, recognize a few familiar words, and immediately jump to the answer choices. Test authors are aware of this common impatience, and they will sometimes prey upon it. For instance, a test author might subtly turn the question into a negative, or he or she might redirect the focus of the question right at the end. The only way to avoid falling into these traps is to read the entirety of the question carefully before reading the answer choices.

3. Looking for Wrong Answers

Long and complicated multiple-choice questions can be intimidating. One way to simplify a difficult multiple-choice question is to eliminate all of the answer choices that are clearly wrong. In most sets of answers, there will be at least one selection that can be dismissed right away. If the test is administered on paper, the test taker could draw a line through it to indicate that it may be ignored; otherwise, the test taker will have to perform this operation mentally or on scratch paper. In either case, once the obviously incorrect answers have been eliminated, the remaining choices may be considered. Sometimes identifying the clearly wrong answers will give the test taker some information about the correct answer. For instance, if one of the remaining answer choices is a direct opposite of one of the eliminated answer choices, it may well be the correct answer. The opposite of obviously wrong is obviously right! Of course, this is not always the case. Some answers are obviously incorrect simply because they are irrelevant to the question being asked. Still, identifying and eliminating some incorrect answer choices is a good way to simplify a multiple-choice question.

4. Don't Overanalyze

Anxious test takers often overanalyze questions. When you are nervous, your brain will often run wild, causing you to make associations and discover clues that don't actually exist. If you feel that this may be a problem for you, do whatever you can to slow down during the test. Try taking a deep breath or counting to ten. As you read and consider the question, restrict yourself to the particular words used by the author. Avoid thought tangents about what the author *really* meant, or what he or she was *trying* to say. The only things that matter on a multiple-choice test are the words that are actually in the question. You must avoid reading too much into a multiple-choice question, or supposing that the writer meant something other than what he or she wrote.

5. No Need for Panic

It is wise to learn as many strategies as possible before taking a multiple-choice test, but it is likely that you will come across a few questions for which you simply don't know the answer. In this situation, avoid panicking. Because most multiple-choice tests include dozens of questions, the relative value of a single wrong answer is small. As much as possible, you should compartmentalize each question on a multiple-choice test. In other words, you should not allow your feelings about one question to affect your success on the others. When you find a question that you either don't understand or don't know how to answer, just take a deep breath and do your best. Read the entire question slowly and carefully. Try rephrasing the question a couple of different ways. Then, read all of the answer choices carefully. After eliminating obviously wrong answers, make a selection and move on to the next question.

6. Confusing Answer Choices

When working on a difficult multiple-choice question, there may be a tendency to focus on the answer choices that are the easiest to understand. Many people, whether consciously or not, gravitate to the answer choices that require the least concentration, knowledge, and memory. This is a mistake. When you come across an answer choice that is confusing, you should give it extra attention. A question might be confusing because you do not know the subject matter to which it refers. If this is the case, don't eliminate the answer before you have affirmatively settled on another. When you come across an answer choice of this type, set it aside as you look at the remaining choices. If you can confidently assert that one of the other choices is correct, you can leave the confusing answer aside. Otherwise, you will need to take a moment to try to better understand the confusing answer choice. Rephrasing is one way to tease out the sense of a confusing answer choice.

7. Your First Instinct

Many people struggle with multiple-choice tests because they overthink the questions. If you have studied sufficiently for the test, you should be prepared to trust your first instinct once you have carefully and completely read the question and all of the answer choices. There is a great deal of research suggesting that the mind can come to the correct conclusion very quickly once it has obtained all of the relevant information. At times, it may seem to you as if your intuition is working faster even than your reasoning mind. This may in fact be true. The knowledge you obtain while studying may be retrieved from your subconscious before you have a chance to work out the associations that support it. Verify your instinct by working out the reasons that it should be trusted.

8. Key Words

Many test takers struggle with multiple-choice questions because they have poor reading comprehension skills. Quickly reading and understanding a multiple-choice question requires a mixture of skill and experience. To help with this, try jotting down a few key words and phrases on a piece of scrap paper. Doing this concentrates the process of reading and forces the mind to weigh the relative importance of the question's parts. In selecting words and phrases to write down, the test taker thinks about the question more deeply and carefully. This is especially true for multiple-choice questions that are preceded by a long prompt.

9. Subtle Negatives

One of the oldest tricks in the multiple-choice test writer's book is to subtly reverse the meaning of a question with a word like *not* or *except*. If you are not paying attention to each word in the question, you can easily be led astray by this trick. For instance, a common question format is, "Which of the following is...?" Obviously, if the question instead is, "Which of the following is not...?," then the answer will be quite different. Even worse, the test makers are aware of the potential for this mistake and will include one answer choice that would be correct if the question were not negated or reversed. A test taker who misses the reversal will find what he or she believes to be a correct answer and will be so confident that he or she will fail to reread the question and discover the original error. The only way to avoid this is to practice a wide variety of multiple-choice questions and to pay close attention to each and every word.

10. Reading Every Answer Choice

It may seem obvious, but you should always read every one of the answer choices! Too many test takers fall into the habit of scanning the question and assuming that they understand the question because they recognize a few key words. From there, they pick the first answer choice that answers the question they believe they have read. Test takers who read all of the answer choices might discover that one of the latter answer choices is actually *more* correct. Moreover, reading all of the answer choices can remind you of facts related to the question that can help you arrive at the correct answer. Sometimes, a misstatement or incorrect detail in one of the latter answer choices will trigger your memory of the subject and will enable you to find the right answer. Failing to read all of the answer choices is like not reading all of the items on a restaurant menu: you might miss out on the perfect choice.

11. Spot the Hedges

One of the keys to success on multiple-choice tests is paying close attention to every word. This is never truer than with words like almost, most, some, and sometimes. These words are called "hedges" because they indicate that a statement is not totally true or not true in every place and time. An absolute statement will contain no hedges, but in many subjects, the answers are not always straightforward or absolute. There are always exceptions to the rules in these subjects. For this reason, you should favor those multiple-choice questions that contain hedging language. The presence of qualifying words indicates that the author is taking special care with his or her words, which is certainly important when composing the right answer. After all, there are many ways to be wrong, but there is only one way to be right! For this reason, it is wise to avoid answers that are absolute when taking a multiple-choice test. An absolute answer is one that says things are either all one way or all another. They often include words like *every*, *always*, *best*, and *never*. If you are taking a multiple-choice test in a subject that doesn't lend itself to absolute answers, be on your guard if you see any of these words.

12. Long Answers

In many subject areas, the answers are not simple. As already mentioned, the right answer often requires hedges. Another common feature of the answers to a complex or subjective question are qualifying clauses, which are groups of words that subtly modify the meaning of the sentence. If the question or answer choice describes a rule to which there are exceptions or the subject matter is complicated, ambiguous, or confusing, the correct answer will require many words in order to be expressed clearly and accurately. In essence, you should not be deterred by answer choices that seem excessively long. Oftentimes, the author of the text will not be able to write the correct answer without

offering some qualifications and modifications. Your job is to read the answer choices thoroughly and completely and to select the one that most accurately and precisely answers the question.

13. Restating to Understand

Sometimes, a question on a multiple-choice test is difficult not because of what it asks but because of how it is written. If this is the case, restate the question or answer choice in different words. This process serves a couple of important purposes. First, it forces you to concentrate on the core of the question. In order to rephrase the question accurately, you have to understand it well. Rephrasing the question will concentrate your mind on the key words and ideas. Second, it will present the information to your mind in a fresh way. This process may trigger your memory and render some useful scrap of information picked up while studying.

14. True Statements

Sometimes an answer choice will be true in itself, but it does not answer the question. This is one of the main reasons why it is essential to read the question carefully and completely before proceeding to the answer choices. Too often, test takers skip ahead to the answer choices and look for true statements. Having found one of these, they are content to select it without reference to the question above. Obviously, this provides an easy way for test makers to play tricks. The savvy test taker will always read the entire question before turning to the answer choices. Then, having settled on a correct answer choice, he or she will refer to the original question and ensure that the selected answer is relevant. The mistake of choosing a correct-but-irrelevant answer choice is especially common on questions related to specific pieces of objective knowledge. A prepared test taker will have a wealth of factual knowledge at his or her disposal, and should not be careless in its application.

15. No Patterns

One of the more dangerous ideas that circulates about multiple-choice tests is that the correct answers tend to fall into patterns. These erroneous ideas range from a belief that B and C are the most common right answers, to the idea that an unprepared test-taker should answer "A-B-A-C-A-D-A-B-A." It cannot be emphasized enough that pattern-seeking of this type is exactly the WRONG way to approach a multiple-choice test. To begin with, it is highly unlikely that the test maker will plot the correct answers according to some predetermined pattern. The questions are scrambled and delivered in a random order. Furthermore, even if the test maker was following a pattern in the assignation of correct answers, there is no reason why the test taker would know which pattern he or she was using. Any attempt to discern a pattern in the answer choices is a waste of time and a distraction from the real work of taking the test. A test taker would be much better served by extra preparation before the test than by reliance on a pattern in the answers.

FREE DVD OFFER

Don't forget that doing well on your exam includes both understanding the test content and understanding how to use what you know to do well on the test. We offer a completely FREE Test Taking Tips DVD that covers world class test taking tips that you can use to be even more successful when you are taking your test.

All that we ask is that you email us your feedback about your study guide. To get your **FREE Test Taking Tips DVD**, email freedvd@studyguideteam.com with "FREE DVD" in the subject line and the following information in the body of the email:

- The title of your study guide.
- Your product rating on a scale of 1-5, with 5 being the highest rating.
- Your feedback about the study guide. What did you think of it?
- Your full name and shipping address to send your free DVD.

Introduction to the ACCUPLACER Exam

Function of the Test

ACCUPLACER is an adaptive, computerized test offered by the College Board and used by some colleges and high schools to determine placement of students in programs appropriate to the students' skill level. The test is offered nation-wide, at any college or high school that chooses to use it. Test-takers are almost always students of schools that use the ACCUPLACER in their course selection and placement efforts. Schools can also use the ACCUPLACER to assess students' skill levels and identify specific areas in which the students need improvement. Scores are generally used only by the college or high school the student is already attending for placement and instruction at that school.

According to the College Board, more than 1,500 secondary and post-secondary institutions are currently using the ACCUPLACER. Over 7.5 million ACCUPLACER tests are administered in a typical year. The College Board recommends that ACCUPLACER scores be used in conjunction with other variables including high school GPA, the number of years a student has taken coursework in a particular subject area, other test scores such as the SAT or ACT, as well as non-cognitive information such as motivation, family support, and time management skills, in order to place students in courses of appropriate difficulty.

Test Administration

ACCUPLACER is offered by computer, usually by the school that wants to use its results, whenever a student makes an appointment with the school to take it. In cases where a student is not able to take the test at the student's school, arrangements can sometimes be made to take the test at a more convenient location on another school's campus.

Fees for taking the ACCUPLACER vary from school to school. Students sometimes do not pay a fee at all; rather, the school pays the College Board for the right to administer the test and then students take it for free. Students may retake the ACCUPLACER at the discretion of the school administering the test and using its results. Students with documented disabilities can make arrangements, through the test center offering the ACCUPLACER, to take the test with appropriate accommodations, including the potential availability of a written version of the test.

Test Format

The core of the ACCUPLACER is five multiple-choice subject area tests: Arithmetic; Quantitative Reasoning, Algebra, and Statistics; Advanced Algebra and Functions, Reading, and Writing. The Writing test has twenty-five questions. It contains one literary passages with five associated questions, and four informational passages that each have five associated questions. A college may ask a student to test in any or all of the five subject areas, depending on the student's and the school's needs. There is also a sixth section, called the WritePlacer, in which students must write a brief essay. Finally, there are four multiple-choice ESL sections of the ACCUPLACER, which schools may ask students, for whom English is a second language, to take.

The multiple-choice sections of the ACCUPLACER are administered by computer. There is no time-limit, so test takers should feel free to work at their own pace. A typical test section might take around 30 minutes to complete. The test is adaptive, meaning it adjusts the difficulty of each question based on

the student's success on the previous questions. The more questions a test-taker gets right, the harder succeeding questions will be and vice-versa. This allows the test to more readily determine test-takers' ability levels without wasting time on questions far above or far below what the test-takers can answer.

Section	Questions	Description
Arithmetic	20	Basic arithmetic and problem solving
Quantitative Reasoning, Algebra, and Statistics	20	Basic algebra, geometry, probability, statistics
Advanced Algebra and Functions	20	Advanced algebra, functions, trigonometry
Reading	20	Information and ideas, rhetoric, synthesis, vocabulary
Writing	25	Expression of ideas, standard English conventions
WritePlacer	1 essay	Effective written communication
ESL- Language Use	20	English grammar
ESL- Listening	20	Understanding spoken English communication
ESL- Reading Skills	20	Comprehension of short written English passages
ESL- Sentence Meaning	20	Understanding the meaning of English sentences

Scoring

The ACCUPLACER subject area tests are scored on a scale from 20 to 120. Schools are free to use the scores for placement as they see fit, given that the difficulty of coursework varies from school to school. A typical community college might separate scores into tiered groups from 50 to 75, from 75 to 99, and from 100 to 120. Scores are typically generated instantly by the computer, but may not be available from the school administering the test until the school has had a chance to review the scores and use them for placement purposes.

Recent Developments

In September 2016, the College Board made "next-generation" ACCUPLACER tests available to schools. Schools have the option of administering the next-generation tests or the older tests, but not both. The new next-generation tests include redesigned reading, writing, and math content, intended to more effectively help schools place students in classes that match their skill level.

Study Prep Plan for the ACCUPLACER Exam

1 **Schedule -** Use one of our study schedules below or come up with one of your own.

2 **Relax -** Test anxiety can hurt even the best students. There are many ways to reduce stress. Find the one that works best for you.

3 **Execute -** Once you have a good plan in place, be sure to stick to it.

Sample Study Plans

One Week Study Schedule		
Day 1	Arithmetic
Day 2	College-Level Math
Day 3	Elementary Algebra
Day 4	Reading Comprehension
Day 5	Sentence Skills
Day 6	WriterPlacer
Day 7	Take Your Exam!

Two Week Study Schedule			
Day 1	Arithmetic	Day 8	Practice Questions
Day 2	Practice Questions	Day 9	Sentence Skills
Day 3	College-Level Math	Day 10	Practice Questions
Day 4	Practice Questions	Day 11	WriterPlacer
Day 5	Elementary Algebra	Day 12	Essay Prompt
Day 6	Practice Questions	Day 13	Review Answer Explanations
Day 7	Reading Comprehension	Day 14	Take Your Exam!

One Month Study Schedule						
Day 1	Addition/ Subtraction	Day 11	Solving Linear Equations	Day 21	Predicates
Day 2	Multiplication/ Division	Day 12	Geometric Reasoning and Graphing	Day 22	Beware of Simplicity
Day 3	Recognizing Equivalent Fractions	Day 13	Practice Questions	Day 23	Practice Questions
Day 4	Practice Questions	Day 14	The Purpose of a Passage	Day 24	Brainstorming
Day 5	Simplifying Rational Algebraic Expressions	Day 15	Types of Passages	Day 25	Considering Opposing Viewpoints
Day 6	Manipulating Roots and Exponents	Day 16	Point of View	Day 26	Moving from Brainstorming to Planning
Day 7	Equation Systems	Day 17	Critical Thinking Skills	Day 27	Parts of the Essay
Day 8	Practice Questions	Day 18	Practice Questions	Day 28	Practice Prompt
Day 9	Evaluation of Simple Formulas	Day 19	Sentence Correction	Day 29	Review Answer Explanations
Day 10	Multiplying and Dividing Monomials	Day 20	Subjects	Day 30	Take Your Exam!

Arithmetic

Whole Number Operations

Addition

Addition is the combination of two numbers so their quantities are added together cumulatively. The sign for an addition operation is the + symbol. For example, 9 + 6 = 15. The 9 and 6 combine to achieve a cumulative value, called a **sum**.

Addition holds the commutative property, which means that the order of the numbers in an addition equation can be switched without altering the result. The formula for the commutative property is a + b = b + a. The following examples can demonstrate how the commutative property works:

$$7 = 3 + 4 = 4 + 3 = 7$$

$$20 = 12 + 8 = 8 + 12 = 20$$

Addition also holds the **associative property**, which means that the grouping of numbers does not matter in an addition problem. In other words, the presence or absence of parentheses is irrelevant. The formula for the associative property is (a + b) + c = a + (b + c). Here are some examples of the associative property at work:

$$30 = (6 + 14) + 10 = 6 + (14 + 10) = 30$$

$$35 = 8 + (2 + 25) = (8 + 2) + 25 = 35$$

Subtraction

Subtraction is taking away one number from another, so their quantities are reduced. The sign designating a subtraction operation is the − symbol, and the result is called the **difference.** For example, 9 - 6 = 3. The number *6* detracts from the number *9* to reach the difference *3*.

Unlike addition, subtraction follows neither the commutative nor associative properties. The order and grouping in subtraction impact the result.

$$15 = 22 - 7 \neq 7 - 22 = -15$$

$$3 = (10 - 5) - 2 \neq 10 - (5 - 2) = 7$$

When working through subtraction problems involving larger numbers, it's necessary to regroup the numbers. The following practice problem uses regrouping:

$$\begin{array}{r} 3\ 2\ 5 \\ -\ 7\ 7 \\ \hline \end{array}$$

Here, it is clear that the ones and tens columns for 77 are greater than the ones and tens columns for 325. To subtract this number, one needs to borrow from the tens and hundreds columns. When borrowing from a column, subtracting 1 from the lender column will add 10 to the borrower column:

$$\begin{array}{r} 3\text{-}1\quad 10+2\text{-}1\quad 10+5 \\ -\qquad\qquad 7\qquad\quad 7 \\ \hline \end{array} = \begin{array}{r} 2\quad 11\quad 15 \\ -\qquad 7\quad 7 \\ \hline 2\quad 4\quad 8 \end{array}$$

After ensuring that each digit in the top row is greater than the digit in the corresponding bottom row, subtraction can proceed as normal, and the answer is found to be 248.

Multiplication

Multiplication involves adding together multiple copies of a number. It is indicated by an $\times$ symbol or a number immediately outside of a parenthesis. For example:

$$5(8 - 2)$$

The two numbers being multiplied together are called **factors**, and their result is called a **product**. For example, $9 \times 6 = 54$. This can be shown alternatively by expansion of either the 9 or the 6:

$$9 \times 6 = 9 + 9 + 9 + 9 + 9 + 9 = 54$$

$$9 \times 6 = 6 + 6 + 6 + 6 + 6 + 6 + 6 + 6 + 6 = 54$$

Like addition, multiplication holds the commutative and associative properties:

$$115 = 23 \times 5 = 5 \times 23 = 115$$

$$84 = 3 \times (7 \times 4) = (3 \times 7) \times 4 = 84$$

Multiplication also follows the **distributive property**, which allows the multiplication to be distributed through parentheses. The formula for distribution is $a \times (b + c) = ab + ac$. This is clear after the examples:

$$45 = 5 \times 9 = 5(3 + 6) = (5 \times 3) + (5 \times 6) = 15 + 30 = 45$$

$$20 = 4 \times 5 = 4(10 - 5) = (4 \times 10) - (4 \times 5) = 40 - 20 = 20$$

Multiplication becomes slightly more complicated when multiplying numbers with decimals. The easiest way to answer these problems is to ignore the decimals and multiply as if they were whole numbers. After multiplying the factors, a decimal gets placed in the product. The placement of the decimal is determined by taking the cumulative number of decimal places in the factors.

For example:

$$0.7$$
$$\underline{\times\ 3}$$
$$2.1$$

$$2.6$$
$$\underline{\times\ \ 4.2}$$
$$10.92$$

$$1.5$$
$$\underline{\times 6.4}$$
$$9.60$$

Starting with the first example, the first step is to ignore the decimal and multiply the numbers as though they were whole numbers, which results in a product of 21. The next step is to count the number of digits that follow a decimal (one, in this case). Finally, the decimal place gets moved that many positions to the left, because the factors have only one decimal place. The second example works the same way, except that there are two total decimal places in the factors, so the product's decimal is moved two places over. In the third example, the decimal should be moved over two digits, but the digit zero is no longer needed, so it is erased, and the final answer is 9.6.

Division

Division and multiplication are inverses of each other in the same way that addition and subtraction are opposites. The signs designating the division operation are the ÷ and / symbols. In division, the second number divides into the first.

The number before the division sign is called the **dividend** or, if expressed as a fraction, the **numerator.** For example, in $a \div b$, a is the dividend, while in $\frac{a}{b}$, a is the numerator.

The number after the division sign is called the **divisor** or, if expressed as a fraction, the **denominator.** For example, in $a \div b$, b is the divisor, while in $\frac{a}{b}$, b is the denominator.

Like subtraction, division doesn't follow the commutative property, as it matters which number comes before the division sign, and division doesn't follow the associative or distributive properties for the same reason. For example:

$$\frac{3}{2} = 9 \div 6 \neq 6 \div 9 = \frac{2}{3}$$

$$2 = 10 \div 5 = (30 \div 3) \div 5 \neq 30 \div (3 \div 5) = 30 \div \frac{3}{5} = 50$$

$$25 = 20 + 5 = (40 \div 2) + (40 \div 8) \neq 40 \div (2 + 8) = 40 \div 10 = 4$$

If a divisor doesn't divide into a dividend an integer number of times, whatever is left over is termed the **remainder**. The remainder can be further divided out into decimal form by using long division; however, this doesn't always give a **quotient** with a finite number of decimal places, so the remainder can also be expressed as a fraction over the original divisor.

Division with decimals is similar to multiplication with decimals in that when dividing a decimal by a whole number, one should ignore the decimal and divide as if it was a whole number.

Upon finding the answer, or quotient, the decimal point is inserted at the decimal place equal to that in the dividend.

$$15.75 \div 3 = 5.25$$

When the divisor is a decimal number, both the divisor and dividend get multiplied by 10. This process is repeated until the divisor is a whole number, then one needs to complete the division operation as described above.

$$17.5 \div 2.5 = 175 \div 25 = 7$$

Order of Operations

When solving equations with multiple operations, special rules apply. These rules are known as the **Order of Operations**. The order is as follows: Parentheses, Exponents, Multiplication and Division from left to right, and Addition and Subtraction from left to right. A popular mnemonic device to help remember the order is Please Excuse My Dear Aunt Sally (PEMDAS).

Evaluate the following two problems to understand the Order of Operations:

1) $4 + (3 \times 2)^2 \div 4$

First, solve the operation within the parentheses: $4 + 6^2 \div 4$.
Second, solve the exponent: $4 + 36 \div 4$.
Third, solve the division operation: $4 + 9$.
Fourth, finish the operation with addition for the answer, 13.

2) $2 \times (6 + 3) \div (2 + 1)^2$

$2 \times 9 \div (3)^2$
$2 \times 9 \div 9$
$18 \div 9$
2

Estimation and Rounding

Estimation is finding a value that is close to a solution but is not the exact answer. For example, if there are values in the thousands to be multiplied, then each value can be estimated to the nearest thousand and the calculation performed. This value provides an approximate solution that can be determined very quickly.

When estimating, it's often convenient to **round** a number, which means to give an approximate figure to make it easier to compare amounts or perform mental math. Round up when the digit is 5 or more. The digit used to determine the rounding, and all subsequent digits, become 0, and the selected place value is increased by 1. Here are some examples:

75 rounded to the nearest ten is 80
380 rounded to the nearest hundred is 400
22.697 rounded to the nearest hundredth is 22.70

Round down when rounding on any digit that is below 5. The rounded digit, and all subsequent digits, becomes 0, and the preceding digit stays the same. Here are some examples:

92 rounded to the nearest ten is 90

839 rounded to the nearest hundred is 800

22.643 rounded to the nearest hundredth is 22.64

The same estimation strategies and techniques used when working with standard math problems can be employed when working with real-life situations. Estimation is frequently used in calculations involving money, such as for determining if one has enough money for a purchase, how much one needs to save weekly to buy a desired product, or how much a restaurant bill will sum to.

Applying Operations to Real-World Contexts

Addition and subtraction are **inverse operations**. Adding a number and then subtracting the same number will cancel each other out, resulting in the original number, and vice versa. For example, $8 + 7 - 7 = 8$ and $137 - 100 + 100 = 137$. Similarly, multiplication and division are inverse operations. Therefore, multiplying by a number and then dividing by the same number results in the original number, and vice versa. For example, $8 \times 2 \div 2 = 8$ and $12 \div 4 \times 4 = 12$. Inverse operations are used to work backwards to solve problems. In the case that 7 and a number add to 18, the inverse operation of subtraction is used to find the unknown value ($18 - 7 = 11$). If a school's entire 4th grade was divided evenly into 3 classes each with 22 students, the inverse operation of multiplication is used to determine the total students in the grade ($22 \times 3 = 66$). Additional scenarios involving inverse operations are included in the tables below.

There are a variety of real-world situations in which one or more of the operators is used to solve a problem. The tables below display the most common scenarios.

Addition & Subtraction

	Unknown Result	Unknown Change	Unknown Start
Adding to	5 students were in class. 4 more students arrived. How many students are in class? $5 + 4 =?$	8 students were in class. More students arrived late. There are now 18 students in class. How many students arrived late? $8+? = 18$ Solved by inverse operations $18- 8 =?$	Some students were in class early. 11 more students arrived. There are now 17 students in class. How many students were in class early? $? +11 = 17$ Solved by inverse operations $17- 11 =?$
Taking from	15 students were in class. 5 students left class. How many students are in class now? $15- 5 =?$	12 students were in class. Some students left class. There are now 8 students in class. How many students left class? $12-? = 8$ Solved by inverse operations $8+? = 12 \rightarrow 12-8 =?$	Some students were in class. 3 students left class. Then there were 13 students in class. How many students were in class before? $?- 3 = 13$ Solved by inverse operations $13 + 3 =?$

	Unknown Total	Unknown Addends (Both)	Unknown Addends (One)
Putting together/ taking apart	The homework assignment is 10 addition problems and 8 subtraction problems. How many problems are in the homework assignment? $10 + 8 =?$	Bobby has $9. How much can Bobby spend on candy and how much can Bobby spend on toys? $9 =? +?$	Bobby has 12 pairs of pants. 5 pairs of pants are shorts, and the rest are long. How many pairs of long pants does he have? $12 = 5+?$ Solved by inverse operations $12- 5 =?$

	Unknown Difference	Unknown Larger Value	Unknown Smaller Value
Comparing	Bobby has 5 toys. Tommy has 8 toys. How many more toys does Tommy have than Bobby? $5+?=8$ Solved by inverse operations $8-5=?$ Bobby has $6. Tommy has $10. How many fewer dollars does Bobby have than Tommy? $10-6=?$	Tommy has 2 more toys than Bobby. Bobby has 4 toys. How many toys does Tommy have? $2+4=?$ Bobby has 3 fewer dollars than Tommy. Bobby has $8. How many dollars does Tommy have? $?-3=8$ Solved by inverse operations $8+3=?$	Tommy has 6 more toys than Bobby. Tommy has 10 toys. How many toys does Bobby have? $?+6=10$ Solved by inverse operations $10-6=?$ Bobby has $5 less than Tommy. Tommy has $9. How many dollars does Bobby have? $9-5=?$

Multiplication and Division

	Unknown Product	Unknown Group Size	Unknown Number of Groups
Equal groups	There are 5 students, and each student has 4 pieces of candy. How many pieces of candy are there in all? $5\times4=?$	14 pieces of candy are shared equally by 7 students. How many pieces of candy does each student have? $7\times?=14$ Solved by inverse operations $14\div7=?$	If 18 pieces of candy are to be given out 3 to each student, how many students will get candy? $?\times3=18$ Solved by inverse operations $18\div3=?$

	Unknown Product	**Unknown Factor**	**Unknown Factor**
Arrays	There are 5 rows of students with 3 students in each row. How many students are there? $5 \times 3 =?$	If 16 students are arranged into 4 equal rows, how many students will be in each row? $4 \times ? = 16$ Solved by inverse operations $16 \div 4 =?$	If 24 students are arranged into an array with 6 columns, how many rows are there? $? \times 6 = 24$ Solved by inverse operations $24 \div 6 =?$

	Larger Unknown	**Smaller Unknown**	**Multiplier Unknown**
Comparing	A small popcorn costs $1.50. A large popcorn costs 3 times as much as a small popcorn. How much does a large popcorn cost? $1.50 \times 3 =?$	A large soda costs $6 and that is 2 times as much as a small soda costs. How much does a small soda cost? $2 \times ? = 6$ Solved by inverse operations $6 \div 2 =?$	A large pretzel costs $3 and a small pretzel costs $2. How many times as much does the large pretzel cost as the small pretzel? $? \times 2 = 3$ Solved by inverse operations $3 \div 2 =?$

Fraction Operations

A **fraction** is a part of something that is whole. Items such as apples can be cut into parts to help visualize fractions. If an apple is cut into 2 equal parts, each part represents ½ of the apple. If each half is cut into two parts, the apple now is cut into quarters. Each piece now represents ¼ of the apple. In this example, each part is equal because they all have the same size. Geometric shapes, such as circles and squares, can also be utilized to help visualize the idea of fractions. For example, a circle can be drawn and divided into 6 equal parts:

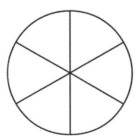

Shading can be used to represent parts of the circle that can be translated into fractions. The top of the fraction, the **numerator**, can represent how many segments are shaded. The bottom of the fraction, the **denominator**, can represent the number of segments that the circle is broken into. A pie is a good

analogy to use in this example. If one piece of the circle is shaded, or one piece of pie is cut out, $^1/_6$ of the object is being referred to. An apple, a pie, or a circle can be utilized in order to compare simple fractions. For example, showing that ½ is larger than ¼ and that ¼ is smaller than $^1/_3$ can be accomplished through shading. A **unit fraction** is a fraction in which the numerator is 1, and the denominator is a positive whole number. It represents one part of a whole—one piece of pie.

Imagine that an apple pie has been baked for a holiday party, and the full pie has eight slices. After the party, there are five slices left. How could the amount of the pie that remains be expressed as a fraction? The numerator is 5 since there are 5 pieces left, and the denominator is 8 since there were eight total slices in the whole pie. Thus, expressed as a fraction, the leftover pie totals $\frac{5}{8}$ of the original amount.

Fractions come in three different varieties: proper fractions, improper fractions, and mixed numbers. **Proper fractions** have a numerator less than the denominator, such as $\frac{3}{8}$, but **improper fractions** have a numerator greater than the denominator, such as $\frac{15}{8}$. **Mixed numbers** combine a whole number with a proper fraction, such as $3\frac{1}{2}$. Any mixed number can be written as an improper fraction by multiplying the integer by the denominator, adding the product to the value of the numerator, and dividing the sum by the original denominator. For example:

$$3\frac{1}{2} = \frac{3 \times 2 + 1}{2} = \frac{7}{2}$$

Whole numbers can also be converted into fractions by placing the whole number as the numerator and making the denominator 1. For example, $3 = \frac{3}{1}$.

The bar in a fraction represents division. Therefore $^6/_5$ is the same as $6 \div 5$. In order to rewrite it as a mixed number, division is performed to obtain $6 \div 5 = 1\,R1$. The remainder is then converted into fraction form. The actual remainder becomes the numerator of a fraction, and the divisor becomes the denominator. Therefore $1\,R1$ is written as $1\frac{1}{5}$, a mixed number. A mixed number can also decompose into the addition of a whole number and a fraction. For example,

$$1\frac{1}{5} = 1 + \frac{1}{5} \text{ and } 4\frac{5}{6} = 4 + \frac{1}{6} + \frac{1}{6} + \frac{1}{6} + \frac{1}{6} + \frac{1}{6}$$

Every fraction can be built from a combination of unit fractions.

One of the most fundamental concepts of fractions is their ability to be manipulated by multiplication or division. This is possible since $\frac{n}{n} = 1$ for any non-zero integer. As a result, multiplying or dividing by $\frac{n}{n}$ will not alter the original fraction since any number multiplied or divided by 1 doesn't change the value of that number. Fractions of the same value are known as equivalent fractions. For example, $\frac{2}{8}, \frac{25}{100}$, and $\frac{40}{160}$ are equivalent, as they all equal $\frac{1}{4}$.

Like fractions, or **equivalent fractions**, are the terms used to describe these fractions that are made up of different numbers but represent the same quantity. For example, the given fractions are $^4/_8$ and $^3/_6$. If a pie was cut into 8 pieces and 4 pieces were removed, half of the pie would remain. Also, if a pie was split into 6 pieces and 3 pieces were eaten, half of the pie would also remain. Therefore, both of the fractions represent half of a pie. These two fractions are referred to as like fractions. **Unlike fractions**

are fractions that are different and cannot be thought of as representing equal quantities. When working with fractions in mathematical expressions, like fractions should be simplified. Both $^4/_8$ and $^3/_6$ can be simplified into $^1/_2$.

Comparing fractions can be completed through the use of a number line. For example, if $^3/_5$ and $^6/_{10}$ need to be compared, each fraction should be plotted on a number line. To plot $^3/_5$, the area from 0 to 1 should be broken into 5 equal segments, and the fraction represents 3 of them. To plot $^6/_{10}$, the area from 0 to 1 should be broken into 10 equal segments and the fraction represents 6 of them.

It can be seen that $\frac{3}{5} = \frac{6}{10}$

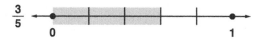

Like fractions are plotted at the same point on a number line. Unit fractions can also be used to compare fractions. For example, if it is known that

$$\frac{4}{5} > \frac{1}{2}$$

and

$$\frac{1}{2} > \frac{4}{10}$$

then it is also known that

$$\frac{4}{5} > \frac{4}{10}$$

Also, converting improper fractions to mixed numbers can be helpful in comparing fractions because the whole number portion of the number is more visible.

Adding and subtracting mixed numbers and fractions can be completed by decomposing fractions into a sum of whole numbers and unit fractions. For example, the given problem is

$$5\frac{3}{7} + 2\frac{1}{7}$$

Decomposing into

$$5 + \frac{1}{7} + \frac{1}{7} + \frac{1}{7} + 2 + \frac{1}{7}$$

This shows that the whole numbers can be added separately from the unit fractions. The answer is:

$$5 + 2 + \frac{1}{7} + \frac{1}{7} + \frac{1}{7} + \frac{1}{7} = 7 + \frac{4}{7} = 7\frac{4}{7}$$

Although many equivalent fractions exist, they are easier to compare and interpret when reduced or simplified. The numerator and denominator of a simple fraction will have no factors in common other than 1. When reducing or simplifying fractions, divide the numerator and denominator by the greatest common factor. A simple strategy is to divide the numerator and denominator by low numbers, like 2, 3, or 5 until arriving at a simple fraction, but the same thing could be achieved by determining the greatest common factor for both the numerator and denominator and dividing each by it. Using the first method is preferable when both the numerator and denominator are even, end in 5, or are obviously a multiple of another number. However, if no numbers seem to work, it will be necessary to factor the numerator and denominator to find the GCF. Let's look at examples:

1) Simplify the fraction $\frac{6}{8}$:

Dividing the numerator and denominator by 2 results in $\frac{3}{4}$, which is a simple fraction.

2) Simplify the fraction $\frac{12}{36}$:

Dividing the numerator and denominator by 2 leaves $\frac{6}{18}$. This isn't a simple fraction, as both the numerator and denominator have factors in common. Diving each by 3 results in $\frac{2}{6}$, but this can be further simplified by dividing by 2 to get $\frac{1}{3}$. This is the simplest fraction, as the numerator is 1. In cases like this, multiple division operations can be avoided by determining the greatest common factor between the numerator and denominator.

3) Simplify the fraction $\frac{18}{54}$ by dividing by the greatest common factor:

First, determine the factors for the numerator and denominator. The factors of 18 are 1, 2, 3, 6, 9, and 18. The factors of 54 are 1, 2, 3, 6, 9, 18, 27, and 54. Thus, the greatest common factor is 18. Dividing $\frac{18}{54}$ by 18 leaves $\frac{1}{3}$, which is the simplest fraction. This method takes slightly more work, but it definitively arrives at the simplest fraction.

Adding and Subtracting Fractions

Adding and subtracting fractions that have the same denominators involves adding or subtracting the numerators. The denominator will stay the same. Therefore, the decomposition process can be made simpler, and the fractions do not have to be broken into unit fractions.

For example, the given problem is:

$$4\frac{7}{8} - 2\frac{6}{8}$$

The answer is found by adding the answers to both

$$4 - 2 \text{ and } \frac{7}{8} - \frac{6}{8}$$

$$2 + \frac{1}{8} = 2\frac{1}{8}$$

A common mistake would be to add the denominators so that

$$\frac{1}{4} + \frac{1}{4} = \frac{1}{8} \text{ or } \frac{2}{8}$$

However, conceptually, it is known that two quarters make a half, so neither one of these are correct.

If two fractions have different denominators, equivalent fractions must be used to add or subtract them. The fractions must be converted into fractions that have common denominators. A **least common denominator** or the product of the two denominators can be used as the common denominator. For example, in the problem $\frac{5}{6} + \frac{2}{3}$, both 6, which is the least common denominator, and 18, which is the product of the denominators, can be used. In order to use 6, $\frac{2}{3}$ must be converted to sixths. A number line can be used to show the equivalent fraction is $\frac{4}{6}$. What happens is that $\frac{2}{3}$ is multiplied times a fractional form of 1 to obtain a denominator of 6. Hence, $\frac{2}{3} \times \frac{2}{2} = \frac{4}{6}$. Therefore, the problem is now

$\frac{5}{6} + \frac{4}{6} = \frac{9}{6}$, which can be simplified into $\frac{3}{2}$. In order to use 18, both fractions must be converted into having 18 as their denominator. $\frac{5}{6}$ would have to be multiplied times $\frac{3}{3}$, and $\frac{2}{3}$ would need to be multiplied times $\frac{6}{6}$. The addition problem would be $\frac{15}{18} + \frac{12}{18} = \frac{27}{18}$, which reduces into $\frac{3}{2}$.

It is always possible to find a common denominator by multiplying the denominators. However, when the denominators are large numbers, this method is unwieldy, especially if the answer must be provided in its simplest form. Thus, it's beneficial to find the **least common denominator** of the fractions—the least common denominator is incidentally also the **least common multiple**.

Once equivalent fractions have been found with common denominators, simply add or subtract the numerators to arrive at the answer:

1) $\frac{1}{2} + \frac{3}{4} = \frac{2}{4} + \frac{3}{4} = \frac{5}{4}$

2) $\frac{3}{12} + \frac{11}{20} = \frac{15}{60} + \frac{33}{60} = \frac{48}{60} = \frac{4}{5}$

3) $\frac{7}{9} - \frac{4}{15} = \frac{35}{45} - \frac{12}{45} = \frac{23}{45}$

4) $\frac{5}{6} - \frac{7}{18} = \frac{15}{18} - \frac{7}{18} = \frac{8}{18} = \frac{4}{9}$

Multiplying and Dividing Fractions

Of the four basic operations that can be performed on fractions, the one that involves the least amount of work is multiplication. To multiply two fractions, simply multiply the numerators together, multiply

the denominators together, and place the products of each as a fraction. Whole numbers and mixed numbers can also be expressed as a fraction, as described above, to multiply with a fraction.

Because multiplication is commutative, multiplying a fraction times a whole number is the same as multiplying a whole number times a fraction. The problem involves adding a fraction a specific number of times. The problem $3 \times \frac{1}{4}$ can be translated into adding the unit fraction 3 times:

$$\frac{1}{4} + \frac{1}{4} + \frac{1}{4} = \frac{3}{4}$$

In the problem $4 \times \frac{2}{5}$, the fraction can be decomposed into $\frac{1}{5} + \frac{1}{5}$ and then added 4 times to obtain $\frac{8}{5}$. Also, both of these answers can be found by just multiplying the whole number times the numerator of the fraction being multiplied.

The whole numbers can be written in fraction form as:

$$\frac{3}{1} \times \frac{1}{4} = \frac{3}{4}$$

$$\frac{4}{1} \times \frac{2}{5} = \frac{8}{5}$$

Multiplying a fraction times a fraction involves multiplying the numerators together separately and the denominators together separately. For example,

$$\frac{3}{8} \times \frac{2}{3} = \frac{3 \times 2}{8 \times 3} = \frac{6}{24}$$

This can then be reduced to $^1/_4$.

Dividing a fraction by a fraction is actually a multiplication problem. It involves flipping the divisor and then multiplying normally. For example,

$$\frac{22}{5} \div \frac{1}{2} = \frac{22}{5} \times \frac{2}{1} = \frac{44}{5}$$

The same procedure can be implemented for division problems involving fractions and whole numbers. The whole number can be rewritten as a fraction over a denominator of 1, and then division can be completed.

A common denominator approach can also be used in dividing fractions. Considering the same problem, $\frac{22}{5} \div \frac{1}{2}$, a common denominator between the two fractions is 10. $\frac{22}{5}$ would be rewritten as $\frac{22}{5} \times \frac{2}{2} = \frac{44}{10}$, and $\frac{1}{2}$ would be rewritten as $\frac{1}{2} \times \frac{5}{5} = \frac{5}{10}$. Dividing both numbers straight across results in:

$$\frac{44}{10} \div \frac{5}{10} = \frac{^{44}/_5}{^{10}/_{10}} = \frac{^{44}/_5}{1} = {^{44}/_5}$$

Many real-world problems will involve the use of fractions. Key words include actual fraction values, such as half, quarter, third, fourth, etc. The best approach to solving word problems involving fractions is to draw a picture or diagram that represents the scenario being discussed, while deciding which type

of operation is necessary in order to solve the problem. A phrase such as "one fourth of 60 pounds of coal" creates a scenario in which multiplication should be used, and the mathematical form of the phrase is $\frac{1}{4} \times 60$.

Decimal Operations

The **decimal system** is a way of writing out numbers that uses ten different numerals: 0, 1, 2, 3, 4, 5, 6, 7, 8, and 9. This is also called a "base ten" or "base 10" system. Other bases are also used. For example, computers work with a base of 2. This means they only use the numerals 0 and 1.

The **decimal place** denotes how far to the right of the decimal point a numeral is. The first digit to the right of the decimal point is in the **tenths** place. The next is the **hundredths**. The third is the *thousandths*.

So, 3.142 has a 1 in the tenths place, a 4 in the hundredths place, and a 2 in the thousandths place.

The **decimal point** is a period used to separate the *ones* place from the *tenths* place when writing out a number as a decimal.

A **decimal number** is a number written out with a decimal point instead of as a fraction, for example, 1.25 instead of $\frac{5}{4}$. Depending on the situation, it can sometimes be easier to work with fractions and sometimes easier to work with decimal numbers.

A decimal number is **terminating** if it stops at some point. It is called **repeating** if it never stops but repeats a pattern over and over. It is important to note that every rational number can be written as a terminating decimal or as a repeating decimal.

Addition with Decimals

To add decimal numbers, each number needs to be lined up by the decimal point in vertical columns. For each number being added, the zeros to the right of the last number need to be filled in so that each of the numbers has the same number of places to the right of the decimal. Then, the columns can be added together. Here is an example of 2.45 + 1.3 + 8.891 written in column form:

$$2.450$$

$$1.300$$

$$+ \ 8.891$$

Zeros have been added in the columns so that each number has the same number of places to the right of the decimal.

Added together, the correct answer is 12.641:

$$2.450$$

$$1.300$$

$$+ \ 8.891$$

12.641

Subtraction with Decimals

Subtracting decimal numbers is the same process as adding decimals. Here is $7.89 - 4.235$ written in column form:

7.890

- 4.235

3.655

A zero has been added in the column so that each number has the same number of places to the right of the decimal.

Multiplication with Decimals

The simplest way to multiply decimals is to calculate the product as if the decimals are not there, then count the number of decimal places in the original problem. Use that total to place the decimal the same number of places over in your answer, counting from right to left. For example, 0.5 x 1.25 can be rewritten and multiplied as 5 x 125, which equals 625. Then the decimal is added three places from the right for .625.

The final answer will have the same number of decimal *points* as the total number of decimal *places* in the problem. The first number has one decimal place, and the second number has two decimal places. Therefore, the final answer will contain three decimal places:

$$0.5 \times 1.25 = 0.625$$

Division with Decimals

Dividing a decimal by a whole number entails using long division first by ignoring the decimal point. Then, the decimal point is moved the number of places given in the problem.

For example, $6.8 \div 4$ can be rewritten as $68 \div 4$, which is 17. There is one non-zero integer to the right of the decimal point, so the final solution would have one decimal place to the right of the solution. In this case, the solution is 1.7.

Dividing a decimal by another decimal requires changing the divisor to a whole number by moving its decimal point. The decimal place of the dividend should be moved by the same number of places as the divisor. Then, the problem is the same as dividing a decimal by a whole number.

For example, $5.72 \div 1.1$ has a divisor with one decimal point in the denominator. The expression can be rewritten as $57.2 \div 11$ by moving each number one decimal place to the right to eliminate the decimal. The long division can be completed as $572 \div 11$ with a result of 52. Since there is one non-zero integer to the right of the decimal point in the problem, the final solution is 5.2.

In another example, $8 \div 0.16$ has a divisor with two decimal points in the denominator. The expression can be rewritten as $800 \div 16$ by moving each number two decimal places to the right to eliminate the decimal in the divisor. The long division can be completed with a result of 50.

Estimation and Rounding with Decimals

Prior to performing operations and calculating the answer to a problem involving addition, subtraction, multiplication, or division, it is helpful to estimate the result. Doing so will enable the test taker to determine whether his or her computed answer is logical within the context of a given problem and prevent careless errors. For example, it is unfortunately common under the pressure of a testing situation for test takers to inadvertently perform the incorrect operation or make a simple calculation error on an otherwise easy math problem. By quickly estimating the answer by eyeballing the numbers, rounding if needed, and performing some simple mental math, test takers can establish an approximate expected outcome before calculating the specific answer. The derived result after computation can then be evaluated by its nearness to the expected answer. This is performed by approximating given values to perform mental math. Numbers should be rounded to the nearest value possible to check the initial results.

As mentioned, sometimes when performing operations such as multiplying numbers, the result can be estimated by rounding. For example, to estimate the value of 11.2×2.01, each number can be rounded to the nearest integer. This will yield a result of 22.

Rounding numbers helps with estimation because it changes the given number to a simpler, although less accurate, number than the exact given number. Rounding allows for easier calculations, which estimate the results of using the exact given number. The accuracy of the estimate and ease of use depends on the place value to which the number is rounded. First, the place value is specified. Then, the digit to its right is looked at. For example, if rounding to the nearest hundreds place, the digit in the tens place is used. If it is a zero, one, 2, 3, or 4, the digit being rounded to is left alone. If it is a 5, 6, 7, 8 or 9, the digit being rounded to is increased by one. All other digits before the decimal point are then changed to zeros, and the digits in decimal places are dropped. If a decimal place is being rounded to, all digits that come after are just dropped.

For example, if 845,231.45 was to be rounded to the nearest thousands place, the answer would be 845,000. The 5 would remain the same due to the 2 in the hundreds place. Also, if 4.567 were to be rounded to the nearest tenths place, the answer would be 4.6. The 5 increased to 6 due to the 6 in the hundredths place, and the rest of the decimal is dropped.

Percent

Percent Problems

Think of percentages as fractions with a denominator of 100. In fact, percentage means "per hundred." The basic percent equation is the following:

$$\frac{is}{of} = \frac{\%}{100}$$

The placement of numbers in the equation depends on what the question asks.

Example 1
Find 40% of 80.

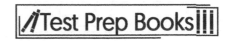

Basically, the problem is asking, "What is 40% of 80?" The 40% is the percent, and 80 is the number to find the percent "of." The equation is:

$$\frac{x}{80} = \frac{40}{100}$$

After cross-multiplying, the problem becomes 100x = 80(40). Solving for x gives the answer: x = 32.

Example 2
What percent of 100 is 20?

The 20 fills in the "is" portion, while 100 fills in the "of." The question asks for the percent, so that will be x, the unknown. The following equation is set up:

$$\frac{20}{100} = \frac{x}{100}$$

Cross-multiplying yields the equation 100x = 20(100). Solving for x gives the answer of 20%.

Example 3
30% of what number is 30?

The following equation uses the clues and numbers in the problem:

$$\frac{30}{x} = \frac{30}{100}$$

Cross-multiplying results in the equation 30(100) = 30x. Solving for x gives the answer x = 100.

Relationships in Numerical Data

In some cases, it is useful to compare numerical data and determine the relationship between values. One of the best ways to mathematically compare two values is to compute the percentage difference between the two values. For example, consider a given music shop that had a net profit of $120,000 in the first year of operation and $185,000 over the second year. Rather than simply finding the net difference between the two years (using subtraction), the business owner may want to know by what percentage his profit increased; in other words, how much his profit in the second year increased relative to his first year. In such cases, the percentage change is desired. The following sections provide some guidance for this process.

Percent Increase/Decrease
Problems dealing with percentages may involve an original value, a change in that value, and a percentage change. A problem will provide two pieces of information and ask to find the third. To do so, this formula is used: $\frac{change}{original\ value} \times 100 =$ percent change. Here's a sample problem:

> Attendance at a baseball stadium has dropped 16% from last year. Last year's average attendance was 40,000. What is this year's average attendance?

Using the formula and information, the change is unknown (x), the original value is 40,000, and the percent change is 16%. The formula can be written as: $\frac{x}{40,000} \times 100 = 16$. When solving for x, it is

determined the change was 6,400. The problem asked for this year's average attendance, so to calculate, the change (6,400) is subtracted from last year's attendance (40,000) to determine this year's average attendance is 33,600.

Percent More Than/Less Than

Percentage problems may give a value and what percent that given value is more than or less than an original unknown value. Here's a sample problem:

A store advertises that all its merchandise has been reduced by 25%. The new price of a pair of shoes is $60. What was the original price?

This problem can be solved by writing a proportion. Two ratios should be written comparing the cost and the percent of the original cost. The new cost is 75% of the original cost (100% - 25%); and the original cost is 100% of the original cost. The unknown original cost can be represented by x. The proportion would be set up as: $\frac{60}{75} = \frac{x}{100}$. Solving the proportion, it is determined the original cost was $80.

Real-World Rate, Percent, and Measurement Problems

A **ratio** compares the size of one group to the size of another. For example, there may be a room with 4 tables and 24 chairs. The ratio of tables to chairs is $4:24$. Such ratios behave like fractions in that both sides of the ratio by the same number can be multiplied or divided. Thus, the ratio 4:24 is the same as the ratio 2:12 and 1:6.

One quantity is **proportional** to another quantity if the first quantity is always some multiple of the second. For instance, the distance travelled in five hours is always five times to the speed as travelled. The distance is proportional to speed in this case.

One quantity is **inversely proportional** to another quantity if the first quantity is equal to some number divided by the second quantity. The time it takes to travel one hundred miles will be given by 100 divided by the speed travelled. The time is inversely proportional to the speed.

When dealing with word problems, there is no fixed series of steps to follow, but there are some general guidelines to use. It is important that the quantity to be found is identified. Then, it can be determined how the given values can be used and manipulated to find the final answer.

Example 1

Jana wants to travel to visit Alice, who lives one hundred and fifty miles away. If she can drive at fifty miles per hour, how long will her trip take?

The quantity to find is the *time* of the trip. The time of a trip is given by the distance to travel divided by the speed to be traveled. The problem determines that the distance is one hundred and fifty miles, while the speed is fifty miles per hour. Thus, 150 divided by 50 is $150 \div 50 = 3$. Because *miles* and *miles per hour* are the units being divided, the miles cancel out. The result is 3 hours.

Example 2

Bernard wishes to paint a wall that measures twenty feet wide by eight feet high. It costs ten cents to paint one square foot. How much money will Bernard need for paint?

The final quantity to compute is the *cost* to paint the wall. This will be ten cents ($0.10) for each square foot of area needed to paint. The area to be painted is unknown, but the dimensions of the wall are given; thus, it can be calculated.

The dimensions of the wall are 20 feet wide and 8 feet high. Since the area of a rectangle is length multiplied by width, the area of the wall is 8 x 20 = 160 square feet. Multiplying 0.1 x 160 yields $16 as the cost of the paint.

Number Comparison and Equivalents

The Position of Numbers Relative to Each Other

Place Value of a Digit

Numbers count in groups of 10. That number is the same throughout the set of natural numbers and whole numbers. It is referred to as working within a base 10 numeration system. Only the numbers from zero to 9 are used to represent any number. The foundation for doing this involves **place value**. Numbers are written side by side. This is to show the amount in each place value.

For place value, let's look at how the number 10 is different from zero to 9. It has two digits instead of just one. The one is in the tens' place, and the zero is in the ones' place. Therefore, there is one group of tens and zero ones. 11 has one 10 and one 1. The introduction of numbers from 11 to 19 should be the next step. Each value within this range of numbers consists of one group of 10 and a specific number of leftover ones. Counting by tens can be practiced once the tens column is understood. This process consists of increasing the number in the tens place by one. For example, counting by 10 starting at 17 would result in the next four values being 27, 37, 47, and 57.

A place value chart can be used for understanding and learning about numbers that have more digits. Here is an example of a place value chart:

	MILLIONS			THOUSANDS			ONES			.	DECIMALS		
billions	hundred millions	ten millions	millions	hundred thousands	ten thousands	thousands	hundreds	tens	ones		tenths	hundredths	thousandths

In the number 1,234, there are 4 ones and 3 tens. The 2 is in the hundreds' place, and the one is in the thousands' place. Note that each group of three digits is separated by a comma. The 2 has a value that is 10 times greater than the 3. Every place to the left has a value 10 times greater than the place to its right. Also, each group of three digits is also known as a *period*. 234 is in the ones' period.

The number 1,234 can be written out as *one-thousand, two hundred thirty-four*. The process of writing out numbers is known as the *decimal system.* It is also based on groups of 10. The place value chart is a helpful tool in using this system. In order to write out a number, it always starts with the digit(s) in the highest period. For example, in the number 23,815,467, the 23 is in highest place and is in the millions' period. The number is read *twenty-three million, eight hundred fifteen thousand, four hundred sixty-seven*. Each period is written separately through the use of commas. Also, no "ands" are used within the number. Another way to think about the number 23,815,467 is through the use of an addition problem. For example:

$$23{,}815{,}467 = 20{,}000{,}000 + 3{,}000{,}000 + 800{,}000 + 10{,}000 + 5{,}000 + 400 + 60 + 7$$

This expression is known as *expanded form*. The actual number 23,815,467 is known as being in **standard form**.

In order to compare whole numbers with many digits, place value can be used. In each number to be compared, it is necessary to find the highest place value in which the numbers differ and to compare the value within that place value. For example, $4{,}523{,}345 < 4{,}532{,}456$ because of the values in the ten thousands place. A similar process can be used for decimals. However, number lines can also be used. Tick marks can be placed within two whole numbers on the number line that represent tenths, hundredths, etc. Each number being compared can then be plotted. The value farthest to the right on the number line is the largest.

Comparing, Classifying, and Ordering Real Numbers

Rational numbers are any number that can be written as a fraction or ratio. Within the set of rational numbers, several subsets exist that are referenced throughout the mathematics topics. **Counting numbers** are the first numbers learned as a child. Counting numbers consist of 1,2,3,4, and so on. **Whole numbers** include all counting numbers and zero (0,1,2,3,4,…). **Integers** include counting numbers, their opposites, and zero (…,-3,-2,-1,0,1,2,3,…). **Rational numbers** are inclusive of integers, fractions, and decimals that terminate, or end (1.7, 0.04213) or repeat ($0.136\bar{5}$).

A **number line** typically consists of integers (…3,2,1,0,-1,-2,-3…), and is used to visually represent the value of a rational number. Each rational number has a distinct position on the line determined by comparing its value with the displayed values on the line. For example, if plotting -1.5 on the number line below, it is necessary to recognize that the value of -1.5 is .5 less than -1 and .5 greater than -2. Therefore, -1.5 is plotted halfway between -1 and -2.

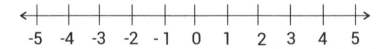

The number system that is used consists of only ten different digits or characters. However, this system is used to represent an infinite number of values. As mentioned, the **place value system** makes this infinite number of values possible. The position in which a digit is written corresponds to a given value. Starting from the decimal point (which is implied, if not physically present), each subsequent place value to the left represents a value greater than the one before it. Conversely, starting from the decimal point, each subsequent place value to the right represents a value less than the one before it.

In accordance with the **base-10 system**, the value of a digit increases by a factor of ten each place it moves to the left. For example, consider the number 7. Moving the digit one place to the left (70), increases its value by a factor of 10 ($7 \times 10 = 70$). Moving the digit two places to the left (700)

increases its value by a factor of 10 twice ($7 \times 10 \times 10 = 700$). Moving the digit three places to the left (7,000) increases its value by a factor of 10 three times ($7 \times 10 \times 10 \times 10 = 7,000$), and so on.

Conversely, the value of a digit decreases by a factor of ten each place it moves to the right. (Note that multiplying by $\frac{1}{10}$ is equivalent to dividing by 10). For example, consider the number 40. Moving the digit one place to the right (4) decreases its value by a factor of 10 ($40 \div 10 = 4$). Moving the digit two places to the right (0.4), decreases its value by a factor of 10 twice ($40 \div 10 \div 10 = 0.4$) or ($40 \times \frac{1}{10} \times \frac{1}{10} = 0.4$). Moving the digit three places to the right (0.04) decreases its value by a factor of 10 three times ($40 \div 10 \div 10 \div 10 = 0.04$) or ($40 \times \frac{1}{10} \times \frac{1}{10} \times \frac{1}{10} = 0.04$), and so on.

Ordering Numbers

A common question type asks to order rational numbers from least to greatest or greatest to least. The numbers will come in a variety of formats, including decimals, percentages, roots, fractions, and whole numbers. These questions test for knowledge of different types of numbers and the ability to determine their respective values.

Before discussing ordering all numbers, let's start with decimals.

To compare decimals and order them by their value, utilize a method similar to that of ordering large numbers.

The main difference is where the comparison will start. Assuming that any numbers to left of the decimal point are equal, the next numbers to be compared are those immediately to the right of the decimal point. If those are equal, then move on to compare the values in the next decimal place to the right.

For example:

Which number is greater, 12.35 or 12.38?

Check that the values to the left of the decimal point are equal:

$$12 = 12$$

Next, compare the values of the decimal place to the right of the decimal:

$$12.3 = 12.3$$

Those are also equal in value.

Finally, compare the value of the numbers in the next decimal place to the right on both numbers:

$$12.3\mathbf{5} \text{ and } 12.3\mathbf{8}$$

Here the 5 is less than the 8, so the final way to express this inequality is:

$$12.35 < 12.38$$

Comparing decimals is regularly exemplified with money because the "cents" portion of money ends in the hundredths place. When paying for gasoline or meals in restaurants, and even in bank accounts, if enough errors are made when calculating numbers to the hundredths place, they can add up to dollars and larger amounts of money over time.

Now that decimal ordering has been explained, let's expand and consider all real numbers. Whether the question asks to order the numbers from greatest to least or least to greatest, the crux of the question is the same—convert the numbers into a common format. Generally, it's easiest to write the numbers as whole numbers and decimals so they can be placed on a number line. Follow these examples to understand this strategy.

1) Order the following rational numbers from greatest to least:

$$\sqrt{36}, 0.65, 78\%, \frac{3}{4}, 7, 90\%, \frac{5}{2}$$

Of the seven numbers, the whole number (7) and decimal (0.65) are already in an accessible form, so concentrate on the other five.

First, the square root of 36 equals 6. (If the test asks for the root of a non-perfect root, determine which two whole numbers the root lies between.) Next, convert the percentages to decimals. A percentage means "per hundred," so this conversion requires moving the decimal point two places to the left, leaving 0.78 and 0.9.

Lastly, evaluate the fractions:

$$\frac{3}{4} = \frac{75}{100} = 0.75 ; \frac{5}{2} = 2\frac{1}{2} = 2.5$$

Now, the only step left is to list the numbers in the request order:

$$7, \sqrt{36}, \frac{5}{2}, 90\%, 78\%, \frac{3}{4}, 0.65$$

2) Order the following rational numbers from least to greatest:

$$2.5, \sqrt{9}, -10.5, 0.853, 175\%, \sqrt{4}, \frac{4}{5}$$

$$\sqrt{9} = 3$$

$$175\% = 1.75$$

$$\sqrt{4} = 2$$

$$\frac{4}{5} = 0.8$$

From least to greatest, the answer is:

$$-10.5, \frac{4}{5}, 0.853, 175\%, \sqrt{4}, 2.5, \sqrt{9}$$

Expressing Numeric Relationships

If a question asks to give words to a mathematical expression and says "equals," then an = sign must be included in the answer. Similarly, "less than or equal to" is expressed by the inequality symbol ≤, and

"greater than or equal" to is expressed as ≥. Furthermore, "less than" is represented by <, and "greater than" is expressed by >.

Equations use the equals sign because the numeric expressions on either side of the symbol (=) are equivalent. In contrast, inequalities compare values or expressions that are unequal. Although not always true, linear equations that include a variable often have just one value for the variable that makes the statement true. Linear inequalities generally have an infinite number of values that make the statement true.

Inequalities are a concise mathematical way to express the relationship between unequal values. More specifically, they describe in what way the values are unequal. A value could be greater than (>); less than (<); greater than or equal to (≥); or less than or equal to (≤) another value. The statement "five times a number added to forty is more than sixty-five" can be expressed as $5x + 40 > 65$. Common words and phrases that express inequalities are:

Symbol	Phrase
<	is under, is below, smaller than, beneath
>	is above, is over, bigger than, exceeds
≤	no more than, at most, maximum
≥	no less than, at least, minimum

Conversions

Decimals and Percentages
Since a percentage is based on "per hundred," decimals and percentages can be converted by multiplying or dividing by 100. Practically speaking, this always amounts to moving the decimal point two places to the right or left, depending on the conversion. To convert a percentage to a decimal, move the decimal point two places to the left and remove the % sign. To convert a decimal to a percentage, move the decimal point two places to the right and add a "%" sign. Here are some examples:

65% = 0.65
0.33 = 33%
0.215 = 21.5%
99.99% = 0.9999
500% = 5.00
7.55 = 755%

Fractions and Percentages
Remember that a percentage is a number per one hundred. So a percentage can be converted to a fraction by making the number in the percentage the numerator and putting 100 as the denominator:

$$43\% = \frac{43}{100}$$

$$97\% = \frac{97}{100}$$

Note that the percent symbol (%) kind of looks like a 0, a 1, and another 0. So think of a percentage like 54% as 54 over 100.

||/Test Prep Books!!!

To convert a fraction to a percent, follow the same logic. If the fraction happens to have 100 in the denominator, you're in luck. Just take the numerator and add a percent symbol:

$$\frac{28}{100} = 28\%$$

Otherwise, divide the numerator by the denominator to get a decimal:

$$\frac{9}{12} = 0.75$$

Then convert the decimal to a percentage:

$$0.75 = 75\%$$

Another option is to make the denominator equal to 100. Be sure to multiply the numerator and the denominator by the same number. For example:

$$\frac{3}{20} \times \frac{5}{5} = \frac{15}{100}$$

$$\frac{15}{100} = 15\%$$

Changing Fractions to Decimals

To change a fraction into a decimal, divide the denominator into the numerator until there are no remainders. There may be repeating decimals, so rounding is often acceptable. A straight line above the repeating portion denotes that the decimal repeats.

Example: Express 4/5 as a decimal.

Set up the division problem.

$$5\overline{)4}$$

5 does not go into 4, so place the decimal and add a zero.

$$5\overline{)4.0}$$

5 goes into 40 eight times. There is no remainder.

$$\begin{array}{r} 0.8 \\ 5\overline{)4.0} \\ -4.0 \\ \hline 0 \end{array}$$

The solution is 0.8.

Example: Express 33 1/3 as a decimal.

Since the whole portion of the number is known, set it aside to calculate the decimal from the fraction portion.

34

Set up the division problem.

$$3\overline{)1}$$

3 does not go into 1, so place the decimal and add zeros. 3 goes into 10 three times.

$$\begin{array}{r} 0.3 \\ 3\overline{)1.0} \end{array}$$

This will repeat with a remainder of 1.

$$\begin{array}{r} 0.333 \\ 3\overline{)1.000} \\ \underline{-9} \\ 10 \\ \underline{-9} \\ 10 \end{array}$$

So, we will place a line over the 3 to denote the repetition. The solution is written $0.\overline{3}$.

Changing Decimals to Fractions

To change decimals to fractions, place the decimal portion of the number, the numerator, over the respective place value, the denominator, then reduce, if possible.

Example: Express 0.25 as a fraction.

This is read as twenty-five hundredths, so put 25 over 100. Then reduce to find the solution.

$$\frac{25}{100} = \frac{1}{4}$$

Example: Express 0.455 as a fraction

This is read as four hundred fifty-five thousandths, so put 455 over 1000. Then reduce to find the solution.

$$\frac{455}{1000} = \frac{91}{200}$$

There are two types of problems that commonly involve percentages. The first is to calculate some percentage of a given quantity, where you convert the percentage to a decimal, and multiply the quantity by that decimal. Secondly, you are given a quantity and told it is a fixed percent of an unknown quantity. In this case, convert to a decimal, then divide the given quantity by that decimal.

Example: What is 30% of 760?

Convert the percent into a useable number. "Of" means to multiply.

$$30\% = 0.30$$

Set up the problem based on the givens, and solve.

$$0.30 \times 760 = 228$$

Example: 8.4 is 20% of what number?

Convert the percent into a useable number.

$$20\% = 0.20$$

The given number is a percent of the answer needed, so divide the given number by this decimal rather than multiplying it.

$$\frac{8.4}{0.20} = 42$$

Distribution of a Quantity into its Fractional Parts

A quantity may be broken into its fractional parts. For example, a toy box holds three types of toys for kids. $\frac{1}{3}$ of the toys are Type A and $\frac{1}{4}$ of the toys are Type B. With that information, how many Type C toys are there?

First, the sum of Type A and Type B must be determined by finding a common denominator to add the fractions. The lowest common multiple is 12, so that is what will be used. The sum is:

$$\frac{1}{3} + \frac{1}{4} = \frac{4}{12} + \frac{3}{12} = \frac{7}{12}$$

This value is subtracted from 1 to find the number of Type C toys. The value is subtracted from 1 because 1 represents a whole. The calculation is:

$$1 - \frac{7}{12} = \frac{12}{12} - \frac{7}{12} = \frac{5}{12}$$

This means that $\frac{5}{12}$ of the toys are Type C. To check the answer, add all fractions together, and the result should be 1.

Practice Questions

1. Which of the following numbers has the greatest value?
    a. 1.4378
    b. 1.07548
    c. 1.43592
    d. 0.89409

2. After a 20% sale discount, Frank purchased a new refrigerator for $850. How much did he save from the original price?
    a. $170
    b. $212.50
    c. $105.75
    d. $200

3. A school has 15 teachers and 20 teaching assistants. They have 200 students. What is the ratio of faculty to students?
    a. 3:20
    b. 4:17
    c. 3:2
    d. 7:40

4. Express the solution to the following problem in decimal form:
$$\frac{3}{5} \times \frac{7}{10} \div \frac{1}{2}$$

    a. 0.042
    b. 84%
    c. 0.84
    d. 0.42

5. If Sarah reads at an average rate of 21 pages in four nights, how long will it take her to read 140 pages?
    a. 6 nights
    b. 26 nights
    c. 8 nights
    d. 27 nights

Answer Explanations

1. A: Compare each numeral after the decimal point to figure out which overall number is greatest. In answers A (1.43785) and C (1.43592), both have the same tenths (4) and hundredths (3). However, the thousandths is greater in answer A (7), so A has the greatest value overall.

2. B: Since $850 is the price *after* a 20% discount, $850 represents 80% of the original price. To determine the original price, set up a proportion with the ratio of the sale price (850) to original price (unknown) equal to the ratio of sale percentage (where x represents the unknown original price):

$$\frac{850}{x} = \frac{80}{100}$$

To solve a proportion, cross multiply the numerators and denominators and set the products equal to each other: $(850)(100) = (80)(x)$. Multiplying each side results in the equation $85{,}000 = 80x$.

To solve for x, divide both sides by 80: $\frac{85{,}000}{80} = \frac{80x}{80}$, resulting in $x = 1062.5$. Remember that x represents the original price. Subtracting the sale price from the original price ($1062.50 - 850) indicates that Frank saved $212.50.

3. D: The total faculty is 15 + 20 = 35. So the ratio is 35:200. Then, divide both of these numbers by 5, since 5 is a common factor to both, with a result of 7:40.

4. C: The first step in solving this problem is expressing the result in fraction form. Separate this problem first by solving the division operation of the last two fractions. When dividing one fraction by another, invert or flip the second fraction and then multiply the numerator and denominator.

$$\frac{7}{10} \times \frac{2}{1} = \frac{14}{10}$$

Next, multiply the first fraction with this value:

$$\frac{3}{5} \times \frac{14}{10} = \frac{42}{50}$$

Decimals are expressions of 1 or 100%, so multiply both the numerator and denominator by 2 to get the fraction as an expression of 100.

$$\frac{42}{50} \times \frac{2}{2} = \frac{84}{100}$$

In decimal form, this would be expressed as 0.84.

5. D: This problem can be solved by setting up a proportion involving the given information and the unknown value. The proportion is:

$$\frac{21\ pages}{4\ nights} = \frac{140\ pages}{x\ nights}$$

Solving the proportion by cross-multiplying, the equation becomes $21x = 4 * 140$, where $x = 26.67$. Since it is not an exact number of nights, the answer is rounded up to 27 nights. Twenty-six nights would not give Sarah enough time.

Quantitative Reasoning, Algebra, and Statistics

Rational Numbers

All real numbers can be separated into two groups: rational and irrational numbers. **Rational numbers** are any numbers that can be written as a fraction, such as $\frac{1}{3}, \frac{7}{4}$, and -25. Alternatively, **irrational numbers** are those that cannot be written as a fraction, such as numbers with never-ending, non-repeating decimal values. Many irrational numbers result from taking roots, such as $\sqrt{2}$ or $\sqrt{3}$. An irrational number may be written as:

$$34.5684952...$$

The ellipsis (...) represents the line of numbers after the decimal that does not repeat and is never-ending.

When rational and irrational numbers interact, there are different types of number outcomes. For example, when adding or multiplying two rational numbers, the result is a rational number. No matter what two fractions are added or multiplied together, the result can always be written as a fraction. The following expression shows two rational numbers multiplied together:

$$\frac{3}{8} \times \frac{4}{7} = \frac{12}{56}$$

The product of these two fractions is another fraction that can be simplified to $\frac{3}{14}$.

As another interaction, rational numbers added to irrational numbers will always result in irrational numbers. No part of any fraction can be added to a never-ending, non-repeating decimal to make a rational number. The same result is true when multiplying a rational and irrational number. Taking a fractional part of a never-ending, non-repeating decimal will always result in another never-ending, non-repeating decimal. An example of the product of rational and irrational numbers is shown in the following expression: $2 \times \sqrt{7}$.

The last type of interaction concerns two irrational numbers, where the sum or product may be rational or irrational depending on the numbers being used. The following expression shows a rational sum from two irrational numbers:

$$\sqrt{3} + \left(6 - \sqrt{3}\right) = 6$$

The product of two irrational numbers can be rational or irrational. A rational result can be seen in the following expression:

$$\sqrt{2} \times \sqrt{8} = \sqrt{2 \times 8} = \sqrt{16} = 4$$

An irrational result can be seen in the following:

$$\sqrt{3} \times \sqrt{2} = \sqrt{6}$$

Structure of the Number System

The mathematical number system is made up of two general types of numbers: real and complex. **Real numbers** are those that are used in normal settings, while **complex numbers** are those composed of both a real number and an imaginary one. Imaginary numbers are the result of taking the square root of -1, and $\sqrt{-1} = i$.

The real number system is often explained using a Venn diagram similar to the one below. After a number has been labeled as a real number, further classification occurs when considering the other groups in this diagram. If a number is a never-ending, non-repeating decimal, it falls in the irrational category. Otherwise, it is rational. More information on these types of numbers is provided in the previous section. Furthermore, if a number does not have a fractional part, it is classified as an integer, such as -2, 75, or zero. Whole numbers are an even smaller group that only includes positive integers and zero. The last group of natural numbers is made up of only positive integers, such as 2, 56, or 12.

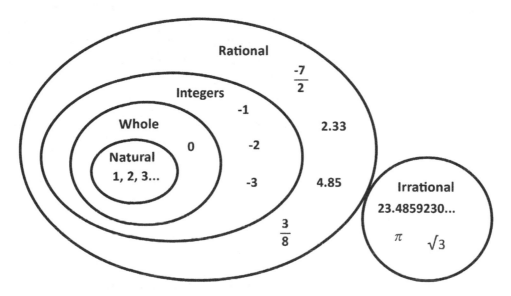

Real numbers can be compared and ordered using the number line. If a number falls to the left on the real number line, it is less than a number on the right. For example, $-2 < 5$ because -2 falls to the left of zero, and 5 falls to the right. Numbers to the left of zero are negative while those to the right are positive.

Complex numbers are made up of the sum of a real number and an imaginary number. Some examples of complex numbers include $6 + 2i, 5 - 7i$, and $-3 + 12i$. Adding and subtracting complex numbers is similar to collecting like terms. The real numbers are added together, and the imaginary numbers are added together. For example, if the problem asks to simplify the expression $6 + 2i - 3 + 7i$, the 6 and (-3) are combined to make 3, and the $2i$ and $7i$ combine to make $9i$. Multiplying and dividing complex numbers is similar to working with exponents. One rule to remember when multiplying is that $i \times i = -1$. For example, if a problem asks to simplify the expression $4i(3 + 7i)$, the $4i$ should be distributed throughout the 3 and the $7i$. This leaves the final expression $12i - 28$. The 28 is negative because $i \times i$ results in a negative number. The last type of operation to consider with complex numbers is the conjugate. The **conjugate** of a complex number is a technique used to change the complex number into a real number. For example, the conjugate of $4 - 3i$ is $4 + 3i$. Multiplying $(4 - 3i)(4 + 3i)$ results in $16 + 12i - 12i + 9$, which has a final answer of $16 + 9 = 25$.

The order of operations—PEMDAS—simplifies longer expressions with real or imaginary numbers. Each operation is listed in the order of how they should be completed in a problem containing more than one operation. Parenthesis can also mean grouping symbols, such as brackets and absolute value. Then, exponents are calculated. Multiplication and division should be completed from left to right, and addition and subtraction should be completed from left to right.

Simplification of another type of expression occurs when radicals are involved. Root is another word for radical. For example, the following expression is a radical that can be simplified: $\sqrt{24x^2}$. First, the number must be factored out to the highest perfect square. Any perfect square can be taken out of a radical. Twenty-four can be factored into 4 and 6, and 4 can be taken out of the radical. $\sqrt{4} = 2$ can be taken out, and 6 stays underneath. If $x > 0$, x can be taken out of the radical because it is a perfect square. The simplified radical is $2x\sqrt{6}$. An approximation can be found using a calculator.

There are also properties of numbers that are true for certain operations. The **commutative** property allows the order of the terms in an expression to change while keeping the same final answer. Both addition and multiplication can be completed in any order and still obtain the same result. However, order does matter in subtraction and division. The **associative** property allows any terms to be "associated" by parenthesis and retain the same final answer. For example:

$$(4 + 3) + 5 = 4 + (3 + 5)$$

Both addition and multiplication are associative; however, subtraction and division do not hold this property. The **distributive** property states that:

$$a(b + c) = ab + ac$$

It is a property that involves both addition and multiplication, and the a is distributed onto each term inside the parentheses.

Absolute Value

The distance that a number is from zero is called its **absolute value**. For example, both 3 and -3 are three spaces from zero. Thus, both -3 and 3 have an absolute value of 3 since they're both three spaces away from zero.

An absolute number is written by placing | | around the number. So, |3| and |−3| both equal 3, as that's their common absolute value.

Ratio and Proportional Relationships

Ratios and Proportional Relationships

Ratios are used to show the relationship between two quantities. The ratio of oranges to apples in the grocery store may be 3 to 2. That means that for every 3 oranges, there are 2 apples. This comparison can be expanded to represent the actual number of oranges and apples. Another example may be the number of boys to girls in a math class. If the ratio of boys to girls is given as 2 to 5, that means there are 2 boys to every 5 girls in the class. Ratios can also be compared if the units in each ratio are the same. The ratio of boys to girls in the math class can be compared to the ratio of boys to girls in a science class by stating which ratio is higher and which is lower.

Rates are used to compare two quantities with different units. **Unit rates** are the simplest form of rate. With unit rates, the denominator in the comparison of two units is one. For example, if someone can type at a rate of 1000 words in 5 minutes, then his or her unit rate for typing is $\frac{1000}{5} = 200$ words in one minute or 200 words per minute. Any rate can be converted into a unit rate by dividing to make the denominator one. 1000 words in 5 minutes has been converted into the unit rate of 200 words per minute.

Ratios and rates can be used together to convert rates into different units. For example, if someone is driving 50 kilometers per hour, that rate can be converted into miles per hour by using a ratio known as the **conversion factor**. Since the given value contains kilometers and the final answer needs to be in miles, the ratio relating miles to kilometers needs to be used. There are 0.62 miles in 1 kilometer. This, written as a ratio and in fraction form, is:

$$\frac{0.62\ miles}{1\ km}$$

To convert 50km/hour into miles per hour, the following conversion needs to be set up:

$$\frac{50\ km}{hour} \times \frac{0.62\ miles}{1\ km} = 31\ miles\ per\ hour$$

The ratio between two similar geometric figures is called the **scale factor**. For example, a problem may depict two similar triangles, A and B. The scale factor from the smaller triangle A to the larger triangle B is given as 2 because the length of the corresponding side of the larger triangle, 16, is twice the corresponding side on the smaller triangle, 8. This scale factor can also be used to find the value of a missing side, x, in triangle A. Since the scale factor from the smaller triangle (A) to larger one (B) is 2, the larger corresponding side in triangle B (given as 25), can be divided by 2 to find the missing side in A ($x = 12.5$). The scale factor can also be represented in the equation $2A = B$ because two times the lengths of A gives the corresponding lengths of B. This is the idea behind similar triangles.

Unit Rate

Unit rate word problems will ask to calculate the rate or quantity of something in a different value. For example, a problem might say that a car drove a certain number of miles in a certain number of minutes and then ask how many miles per hour the car was traveling. These questions involve solving proportions. Consider the following examples:

1) Alexandra made $96 during the first 3 hours of her shift as a temporary worker at a law office. She will continue to earn money at this rate until she finishes in 5 more hours. How much does Alexandra make per hour? How much will Alexandra have made at the end of the day?

This problem can be solved in two ways. The first is to set up a proportion, as the rate of pay is constant. The second is to determine her hourly rate, multiply the 5 hours by that rate, and then add the $96.

To set up a proportion, put the money already earned over the hours already worked on one side of an equation. The other side has x over 8 hours (the total hours worked in the day). It looks like this:

$$\frac{96}{3} = \frac{x}{8}$$

Now, cross-multiply to get $768 = 3x$. To get x, divide by 3, which leaves $x = 256$. Alternatively, as x is the numerator of one of the proportions, multiplying by its denominator will reduce the solution by one step. Thus, Alexandra will make $256 at the end of the day. To calculate her hourly rate, divide the total by 8, giving $32 per hour.

Alternatively, it is possible to figure out the hourly rate by dividing $96 by 3 hours to get $32 per hour. Now her total pay can be figured by multiplying $32 per hour by 8 hours, which comes out to $256.

2) Jonathan is reading a novel. So far, he has read 215 of the 335 total pages. It takes Jonathan 25 minutes to read 10 pages, and the rate is constant. How long does it take Jonathan to read one page? How much longer will it take him to finish the novel? Express the answer in time.

To calculate how long it takes Jonathan to read one page, divide the 25 minutes by 10 pages to determine the page per minute rate. Thus, it takes 2.5 minutes to read one page.

Jonathan must read 120 more pages to complete the novel. (This is calculated by subtracting the pages already read from the total.) Now, multiply his rate per page by the number of pages. Thus, $120 \times 2.5 = 300$. Expressed in time, 300 minutes is equal to 5 hours.

3) At a hotel, $\frac{4}{5}$ of the 120 rooms are booked for Saturday. On Sunday, $\frac{3}{4}$ of the rooms are booked. On which day are more of the rooms booked, and by how many more?

The first step is to calculate the number of rooms booked for each day. Do this by multiplying the fraction of the rooms booked by the total number of rooms.

$$\text{Saturday: } \frac{4}{5} \times 120 = \frac{4}{5} \times \frac{120}{1} = \frac{480}{5} = 96 \text{ rooms}$$

$$\text{Sunday: } \frac{3}{4} \times 120 = \frac{3}{4} \times \frac{120}{1} = \frac{360}{4} = 90 \text{ rooms}$$

Thus, more rooms were booked on Saturday by 6 rooms.

4) In a veterinary hospital, the veterinarian-to-pet ratio is 1:9. The ratio is always constant. If there are 45 pets in the hospital, how many veterinarians are currently in the veterinary hospital?

Set up a proportion to solve for the number of veterinarians: $\frac{1}{9} = \frac{x}{45}$

Cross-multiplying results in $9x = 45$, which works out to 5 veterinarians.

Alternatively, as there are always 9 times as many pets as veterinarians, it is possible to divide the number of pets (45) by 9. This also arrives at the correct answer of 5 veterinarians.

5) At a general practice law firm, 30% of the lawyers work solely on tort cases. If 9 lawyers work solely on tort cases, how many lawyers work at the firm?

First, solve for the total number of lawyers working at the firm, which will be represented here with x. The problem states that 9 lawyers work solely on torts cases, and they make up 30% of the total lawyers at the firm. Thus, 30% multiplied by the total, x, will equal 9. Written as equation, this is:

$$30\% \times x = 9$$

It's easier to deal with the equation after converting the percentage to a decimal, leaving $0.3x = 9$. Thus, $x = \frac{9}{0.3} = 30$ lawyers working at the firm.

6) Xavier was hospitalized with pneumonia. He was originally given 35mg of antibiotics. Later, after his condition continued to worsen, Xavier's dosage was increased to 60mg. What was the percent increase of the antibiotics? Round the percentage to the nearest tenth.

An increase or decrease in percentage can be calculated by dividing the difference in amounts by the original amount and multiplying by 100. Written as an equation, the formula is:

$$\frac{new\ quantity\ -\ old\ quantity}{old\ quantity} \times 100$$

Here, the question states that the dosage was increased from 35mg to 60mg, so these are plugged into the formula to find the percentage increase.

$$\frac{60 - 35}{35} \times 100 = \frac{25}{35} \times 100 = 0.7142 \times 100 = 71.4\%$$

Analyzing Proportional Relationships and Using Them to Solve Real-World Problems

Much like a scale factor can be written using an equation like $2A = B$, a **relationship** is represented by the equation $Y = kX$. X and Y are proportional because as values of X increase, the values of Y also increase. A relationship that is inversely proportional can be represented by the equation $Y = \frac{k}{x}$, where the value of Y decreases as the value of x increases and vice versa.

Proportional reasoning can be used to solve problems involving ratios, percentages, and averages. Ratios can be used in setting up proportions and solving them to find unknowns. For example, if a student completes an average of 10 pages of math homework in 3 nights, how long would it take the student to complete 22 pages? Both ratios can be written as fractions. The second ratio would contain the unknown.

The following proportion represents this problem, where x is the unknown number of nights:

$$\frac{10\ pages}{3\ nights} = \frac{22\ pages}{x\ nights}$$

Solving this proportion entails cross-multiplying and results in the following equation: $10x = 22 \times 3$. Simplifying and solving for x results in the exact solution: $x = 6.6\ nights$. The result would be rounded up to 7 because the homework would actually be completed on the 7th night.

The following problem uses ratios involving percentages:

If 20% of the class is girls and 30 students are in the class, how many girls are in the class?

To set up this problem, it is helpful to use the common proportion:

$$\frac{\%}{100} = \frac{is}{of}$$

Within the proportion, % is the percentage of girls, 100 is the total percentage of the class, *is* is the number of girls, and *of* is the total number of students in the class. Most percentage problems can be written using this language. To solve this problem, the proportion should be set up as $\frac{20}{100} = \frac{x}{30}$, and then solved for x. Cross-multiplying results in the equation $20 \times 30 = 100x$, which results in the solution $x = 6$. There are 6 girls in the class.

Problems involving volume, length, and other units can also be solved using ratios. For example, a problem may ask for the volume of a cone to be found that has a radius, $r = 7m$ and a height, $h = 16m$. Referring to the formulas provided on the test, the volume of a cone is given as: $V = \pi r^2 \frac{h}{3}$, where r is the radius, and h is the height. Plugging $r = 7$ and $h = 16$ into the formula, the following is obtained:

$$V = \pi (7^2)\frac{16}{3}$$

Therefore, volume of the cone is found to be approximately 821m³. Sometimes, answers in different units are sought. If this problem wanted the answer in liters, 821m³ would need to be converted. Using the equivalence statement 1m³ = 1000L, the following ratio would be used to solve for liters:

$$821\text{m}^3 \times \frac{1000L}{1m^3}$$

Cubic meters in the numerator and denominator cancel each other out, and the answer is converted to 821,000 liters, or 8.21×10^5 L.

Other conversions can also be made between different given and final units. If the temperature in a pool is 30°C, what is the temperature of the pool in degrees Fahrenheit? To convert these units, an equation is used relating Celsius to Fahrenheit. The following equation is used:

$$T_{°F} = 1.8T_{°C} + 32$$

Plugging in the given temperature and solving the equation for T yields the result:

$$T_{°F} = 1.8(30) + 32 = 86°F$$

Both units in the metric system and U.S. customary system are widely used.

Here are some more examples of how to solve for proportions:

1) $\frac{75\%}{90\%} = \frac{25\%}{x}$

To solve for x, the fractions must be cross multiplied:

$$(75\%x = 90\% \times 25\%)$$

To make things easier, let's convert the percentages to decimals:

$$(0.9 \times 0.25 = 0.225 = 0.75x)$$

To get rid of x's coefficient, each side must be divided by that same coefficient to get the answer:

$$x = 0.3$$

The question could ask for the answer as a percentage or fraction in lowest terms, which are 30% and $\frac{3}{10}$, respectively.

2) $\frac{x}{12} = \frac{30}{96}$

Cross-multiply: $96x = 30 \times 12$

Multiply: $96x = 360$

Divide: $x = 360 \div 96$

Answer: $x = 3.75$

3) $\frac{0.5}{3} = \frac{x}{6}$

Cross-multiply: $3x = 0.5 \times 6$

Multiply: $3x = 3$

Divide: $x = 3 \div 3$

Answer: $x = 1$

You may have noticed there's a faster way to arrive at the answer. If there is an obvious operation being performed on the proportion, the same operation can be used on the other side of the proportion to solve for x. For example, in the first practice problem, 75% became 25% when divided by 3, and upon doing the same to 90%, the correct answer of 30% would have been found with much less legwork. However, these questions aren't always so intuitive, so it's a good idea to work through the steps, even if the answer seems apparent from the outset.

Exponents

Radicals and Exponents

Exponents are used in mathematics to express a number or variable multiplied by itself a certain number of times. For example, x^3 means x is multiplied by itself three times. In this expression, x is called the **base**, and 3 is the **exponent**. Exponents can be used in more complex problems when they contain fractions and negative numbers.

Fractional exponents can be explained by looking first at the inverse of exponents, which are **roots**. Given the expression x^2, the square root can be taken, $\sqrt{x^2}$, cancelling out the 2 and leaving x by itself, if x is positive. Cancellation occurs because $\sqrt{x}$ can be written with exponents, instead of roots, as $x^{\frac{1}{2}}$. The numerator of 1 is the exponent, and the denominator of 2 is called the root (which is why it's referred to as **square root**). Taking the square root of x^2 is the same as raising it to the $\frac{1}{2}$ power. Written out in mathematical form, it takes the following progression:

$$\sqrt{x^2} = (x^2)^{\frac{1}{2}} = x$$

From properties of exponents, $2 \times \frac{1}{2} = 1$ is the actual exponent of x. Another example can be seen with $x^{\frac{4}{7}}$. The variable x, raised to four-sevenths, is equal to the seventh root of x to the fourth power: $\sqrt[7]{x^4}$. In general,

$$x^{\frac{1}{n}} = \sqrt[n]{x}$$

and

$$x^{\frac{m}{n}} = \sqrt[n]{x^m}$$

Negative exponents also involve fractions. Whereas y^3 can also be rewritten as $\frac{y^3}{1}$, y^{-3} can be rewritten as $\frac{1}{y^3}$. A negative exponent means the exponential expression must be moved to the opposite spot in a fraction to make the exponent positive. If the negative appears in the numerator, it moves to the denominator. If the negative appears in the denominator, it is moved to the numerator. In general, $a^{-n} = \frac{1}{a^n}$, and a^{-n} and a^n are reciprocals.

Take, for example, the following expression:

$$\frac{a^{-4}b^2}{c^{-5}}$$

Since a is raised to the negative fourth power, it can be moved to the denominator. Since c is raised to the negative fifth power, it can be moved to the numerator. The b variable is raised to the positive second power, so it does not move.

The simplified expression is as follows:

$$\frac{b^2 c^5}{a^4}$$

In mathematical expressions containing exponents and other operations, the order of operations must be followed. PEMDAS states that exponents are calculated after any parenthesis and grouping symbols, but before any multiplication, division, addition, and subtraction.

There are a few rules for working with exponents. For any numbers a, b, m, n, the following hold true:

$$a^1 = a$$

$$1^a = 1$$

$$a^0 = 1$$

$$a^m \times a^n = a^{m+n}$$

$$a^m \div a^n = a^{m-n}$$

$$(a^m)^n = a^{m \times n}$$

$$(a \times b)^m = a^m \times b^m$$

$$(a \div b)^m = a^m \div b^m$$

Any number, including a fraction, can be an exponent. The same rules apply.

Manipulating Roots and Exponents

A **root** is a different way to write an exponent when the exponent is the reciprocal of a whole number. We use the **radical** symbol to write this in the following way:

$$\sqrt[n]{a} = a^{\frac{1}{n}}$$

This quantity is called the *n-th root* of *a*. The *n* is called the **index** of the radical.

Note that if the *n*-th root of *a* is multiplied by itself *n* times, the result will just be *a*. If no number *n* is written by the radical, it is assumed that *n* is 2:

$$\sqrt{5} = 5^{\frac{1}{2}}$$

The special case of the 2ⁿᵈ root is called the **square root**, and the third root is called the **cube root**.

A **perfect square** is a whole number that is the square of another whole number. For example, sixteen and 64 are perfect squares because 16 is the square of 4, and 64 is the square of 8.

Scientific Notation

Scientific notation is the conversion of extremely small or large numbers into a format that is easier to comprehend and manipulate. It changes the number into three separate parts: a mathematical sign (+/−), a digit term (known as a **significand**), and an exponential term.

Scientific notation = (+ or -) significand x exponential term

To put a number into scientific notation, one should use the following steps:

- Move the decimal point to after the first non-zero number to find the digit number.
- Count how many places the decimal point was moved in step 1.
- Determine if the exponent is positive or negative.
- Create an exponential term using the information from steps 2 and 3.
- Combine the digit term and exponential term to get scientific notation.

For example, to put 0.0000098 into scientific notation, the decimal should be moved so that it lies between the last two numbers: 000009.8. This creates the digit number:

9.8

Next, the number of places that the decimal point moved is determined; to get between the 9 and the 8, the decimal was moved six places to the right. It may be helpful to remember that a decimal moved to the right creates a negative exponent, and a decimal moved to the left creates a positive exponent. Because the decimal was moved six places to the right, the exponent is negative.

Now, the exponential term can be created by using the base 10 (this is *always* the base in scientific notation) and the number of places moved as the exponent, in this case:

$$10^{-6}$$

Finally, the digit term and the exponential term can be combined as a product. Therefore, the scientific notation for the number 0.0000098 is:

$$9.8 \times 10^{-6}$$

Algebraic Expressions

Algebraic Expressions

Algebraic expressions look similar to equations, but they do not include the equal sign. Algebraic expressions are comprised of numbers, variables, and mathematical operations. Some examples of algebraic expressions are $8x + 7y - 12z$, $3a^2$, and $5x^3 - 4y^4$.

Algebraic expressions consist of variables, numbers, and operations. A **term** of an expression is any combination of numbers and/or variables, and terms are separated by addition and subtraction. For example, the expression $5x^2 - 3xy + 4 - 2$ consists of 4 terms: $5x^2$, -3xy, 4y, and -2. Note that each term includes its given sign (+ or −). The **variable** part of a term is a letter that represents an unknown quantity. The coefficient of a term is the number by which the variable is multiplied. For the term 4y, the variable is y and the coefficient is 4. Terms are identified by the power (or exponent) of its variable.

A number without a variable is referred to as a **constant**. If the variable is to the first power (x^1 or simply x), it is referred to as a **linear** term. A term with a variable to the second power (x^2) is **quadratic** and a term to the third power (x^3) is **cubic**. Consider the expression $x^3 + 3x - 1$. The constant is -1. The linear term is 3x. There is no quadratic term. The cubic term is x^3.

An algebraic expression can also be classified by how many terms exist in the expression. Any like terms should be combined before classifying. A **monomial** is an expression consisting of only one term. Examples of monomials are: 17, 2x, and $-5ab^2$. A **binomial** is an expression consisting of two terms separated by addition or subtraction. Examples include $2x - 4$ and $-3y^2 + 2y$. A **trinomial** consists of 3 terms. For example, $5x^2 - 2x + 1$ is a trinomial.

Algebraic expressions and equations can be used to represent real-life situations and model the behavior of different variables. For example, $2x + 5$ could represent the cost to play games at an arcade. In this case, 5 represents the price of admission to the arcade and 2 represents the cost of each game played. To calculate the total cost, use the number of games played for x, multiply it by 2, and add 5.

Adding and Subtracting Linear Algebraic Expressions

An algebraic expression is simplified by combining like terms. A term is a number, variable, or product of a number, and variables separated by addition and subtraction. For the algebraic expression $3x^2 - 4x + 5 - 5x^2 + x - 3$, the terms are $3x^2$, -4x, 5, $-5x^2$, x, and -3. **Like terms** have the same variables raised to the same powers (exponents). The like terms for the previous example are $3x^2$ and $-5x^2$, -4x and x, 5 and -3. To combine like terms, the coefficients (numerical factor of the term including sign) are added and the variables and their powers are kept the same. Note that if a coefficient is not written, it is an implied coefficient of 1 ($x = 1x$). The previous example will simplify to $-2x^2 - 3x + 2$.

When adding or subtracting algebraic expressions, each expression is written in parenthesis. The negative sign is distributed when necessary, and like terms are combined. Consider the following:

$$\text{add } 2a + 5b - 2 \text{ to } a - 2b + 8c - 4$$

The sum is set as follows:

$$(a - 2b + 8c - 4) + (2a + 5b - 2)$$

In front of each set of parentheses is an implied positive one, which, when distributed, does not change any of the terms. Therefore, the parentheses are dropped and like terms are combined:

$$a - 2b + 8c - 4 + 2a + 5b - 2$$

$$3a + 3b + 8c - 6$$

Consider the following problem:

$$\text{Subtract } 2a + 5b - 2 \text{ from } a - 2b + 8c - 4$$

The difference is set as follows:

$$(a - 2b + 8c - 4) - (2a + 5b - 2)$$

The implied one in front of the first set of parentheses will not change those four terms. However, distributing the implied -1 in front of the second set of parentheses will change the sign of each of those three terms:

$$a - 2b + 8c - 4 - 2a - 5b + 2$$

Combining like terms yields the simplified expression:

$$-a - 7b + 8c - 2$$

Multiplying Algebraic Expressions

The **distributive property** states that multiplying a sum (or difference) by a number produces the same result as multiplying each value in the sum (or difference) by the number and adding (or subtracting) the products. Using mathematical symbols, the distributive property states $a(b + c) = ab + ac$. The expression $4(3 + 2)$ is simplified using the order of operations. Simplifying inside the parenthesis first produces 4×5, which equals 20. The expression $4(3 + 2)$ can also be simplified using the distributive property:

$$4(3 + 2)$$

$$4 \times 3 + 4 \times 2$$

$$12 + 8$$

$$20$$

Consider the following example: $4(3x - 2)$. The expression cannot be simplified inside the parenthesis because $3x$ and -2 are not like terms, and therefore cannot be combined. However, the expression can be simplified by using the distributive property and multiplying each term inside of the parenthesis by the term outside of the parenthesis: $12x - 8$. The resulting equivalent expression contains no like terms, so it cannot be further simplified.

Consider the expression:

$$(3x + 2y + 1) - (5x - 3) + 2(3y + 4)$$

Again, there are no like terms, but the distributive property is used to simplify the expression. Note there is an implied one in front of the first set of parentheses and an implied -1 in front of the second set of parentheses. Distributing the one, -1, and 2 produces:

$$1(3x) + 1(2y) + 1(1) - 1(5x) - 1(-3) + 2(3y) + 2(4)$$

$$3x + 2y + 1 - 5x + 3 + 6y + 8$$

This expression contains like terms that are combined to produce the simplified expression:

$$-2x + 8y + 12$$

Algebraic expressions are tested to be equivalent by choosing values for the variables and evaluating both expressions. For example, $4(3x - 2)$ and $12x - 8$ are tested by substituting 3 for the variable x and calculating to determine if equivalent values result.

Evaluating Algebraic Expressions

To evaluate the expression, the given values for the variables are substituted (or replaced) and the expression is simplified using the order of operations. Parenthesis should be used when substituting. Consider the following: Evaluate $a - 2b + ab$ for $a = 3$ and $b = -1$. To evaluate, any variable a is replaced with 3 and any variable b with -1, producing:

$$(3) - 2(-1) + (3)(-1)$$

Next, the order of operations is used to calculate the value of the expression, which is 2.

Here's another example:

$$\text{Evaluate } a - 2b + ab \text{ for } a = 3 \text{ and } b = -1$$

To evaluate an expression, the given values should be substituted for the variables and simplified using the order of operations. In this case:

$$(3) - 2(-1) + (3)(-1)$$

Parentheses are used when substituting.

Given an algebraic expression, students may be asked to simplify the expression. For example:

$$\text{Simplify } 5x^2 - 10x + 2 - 8x^2 + x - 1.$$

Simplifying algebraic expressions requires combining like terms. A term is a number, variable, or product of a number and variables separated by addition and subtraction. The terms in the above expression are: $5x^2, -10x, 2, -8x^2, x$, and -1. Like terms have the same variables raised to the same powers

(exponents). To combine like terms, the coefficients (numerical factor of the term including sign) are added, while the variables and their powers are kept the same. The example above simplifies to:

$$-3x^2 - 9x + 1$$

Let's try two more.

Evaluate $\frac{1}{2}x^2 - 3, x = 4$.

The first step is to substitute in 4 for x in the expression:

$$\frac{1}{2}(4)^2 - 3$$

Then, the order of operations is used to simplify.

The exponent comes first, $\frac{1}{2}(16) - 3$, then the multiplication $8 - 3$, and then, after subtraction, the solution is 5.

Evaluate $4|5 - x| + 2y, x = 4, y = -3$.

The first step is to substitute 4 in for x and -3 in for y in the expression:

$$4|5 - 4| + 2(-3)$$

Then, the absolute value expression is simplified, which is:

$$|5 - 4| = |1| = 1$$

The expression is $4(1) + 2(-3)$ which can be simplified using the order of operations.

First is the multiplication, $4 + (-6)$; then addition yields an answer of -2.

Creating Algebraic Expressions

A linear expression is a statement about an unknown quantity expressed in mathematical symbols. The statement "five times a number added to forty" can be expressed as $5x + 40$. A linear equation is a statement in which two expressions (at least one containing a variable) are equal to each other. The statement "five times a number added to forty is equal to ten" can be expressed as $5x + 40 = 10$.

Real world scenarios can also be expressed mathematically. Suppose a job pays its employees $300 per week and $40 for each sale made. The weekly pay is represented by the expression $40x + 300$ where x is the number of sales made during the week.

Consider the following scenario: Bob had $20 and Tom had $4. After selling 4 ice cream cones to Bob, Tom has as much money as Bob. The cost of an ice cream cone is an unknown quantity and can be represented by a variable (x). The amount of money Bob has after his purchase is four times the cost of an ice cream cone subtracted from his original $20 → $20 - 4x$. The amount of money Tom has after his sale is four times the cost of an ice cream cone added to his original $4 → $4x + 4$. After the sale, the amount of money that Bob and Tom have are equal → $20 - 4x = 4x + 4$.

When expressing a verbal or written statement mathematically, it is key to understand words or phrases that can be represented with symbols. The following are examples:

Symbol	Phrase
$+$	added to, increased by, sum of, more than
$-$	decreased by, difference between, less than, take away
x	multiplied by, 3 (4, 5 . . .) times as large, product of
$\div$	divided by, quotient of, half (third, etc.) of
$=$	is, the same as, results in, as much as
$x, t, n, etc.$	a number, unknown quantity, value of

Linear Equations

One-Variable Linear Equations and Inequalities

Linear equations and linear inequalities are both comparisons of two algebraic expressions. However, unlike equations in which the expressions are equal, linear inequalities compare expressions that may be unequal. Linear equations typically have one value for the variable that makes the statement true. Linear inequalities generally have an infinite number of values that make the statement true.

When solving a linear equation, the desired result requires determining a numerical value for the unknown variable. If given a linear equation involving addition, subtraction, multiplication, or division, working backwards isolates the variable. Addition and subtraction are inverse operations, as are multiplication and division. Therefore, they can be used to cancel each other out.

For example, solve $4(t - 2) + 2t - 4 = 2(9 - 2t)$

Distributing: $4t - 8 + 2t - 4 = 18 - 4t$

Combining like terms: $6t - 12 = 18 - 4t$

Adding $4t$ to each side to move the variable: $10t - 12 = 18$

Adding 12 to each side to isolate the variable: $10t = 30$

Dividing each side by 10 to isolate the variable: $t = 3$

The answer can be checked by substituting the value for the variable into the original equation, ensuring that both sides calculate to be equal.

Linear inequalities express the relationship between unequal values. More specifically, they describe in what way the values are unequal. A value can be greater than (>), less than (<), greater than or equal to (≥), or less than or equal to (≤) another value. $5x + 40 > 65$ is read as *five times a number added to forty is greater than sixty-five.*

When solving a linear inequality, the solution is the set of all numbers that make the statement true. The inequality $x + 2 \geq 6$ has a solution set of 4 and every number greater than 4 (4.01; 5; 12; 107; etc.). Adding 2 to 4 or any number greater than 4 results in a value that is greater than or equal to 6. Therefore, $x \geq 4$ is the solution set.

To algebraically solve a linear inequality, follow the same steps as those for solving a linear equation. The inequality symbol stays the same for all operations *except* when multiplying or dividing by a negative number. If multiplying or dividing by a negative number while solving an inequality, the relationship reverses (the sign flips). In other words, > switches to < and vice versa. Multiplying or dividing by a positive number does not change the relationship, so the sign stays the same. An example is shown below.

Solve $-2x - 8 \leq 22$

Add 8 to both sides: $-2x \leq 30$

Divide both sides by -2: $x \geq -15$

Solutions of a linear equation or a linear inequality are the values of the variable that make a statement true. In the case of a linear equation, the solution set (list of all possible solutions) typically consists of a single numerical value. To find the solution, the equation is solved by isolating the variable. For example, solving the equation $3x - 7 = -13$ produces the solution $x = -2$. The only value for x which produces a true statement is -2. This can be checked by substituting -2 into the original equation to check that both sides are equal. In this case, $3(-2) - 7 = -13 \rightarrow -13 = -13$; therefore, -2 is a solution.

Although linear equations generally have one solution, this is not always the case. If there is no value for the variable that makes the statement true, there is no solution to the equation. Consider the equation $x + 3 = x - 1$. There is no value for x in which adding 3 to the value produces the same result as subtracting one from the value. Conversely, if any value for the variable makes a true statement, the equation has an infinite number of solutions. Consider the equation:

$$3x + 6 = 3(x + 2)$$

Any number substituted for x will result in a true statement (both sides of the equation are equal).

By manipulating equations like the two above, the variable of the equation will cancel out completely. If the remaining constants express a true statement (ex. $6 = 6$), then all real numbers are solutions to the equation. If the constants left express a false statement (ex. $3 = -1$), then no solution exists for the equation.

Solving a linear inequality requires all values that make the statement true to be determined. For example, solving $3x - 7 \geq -13$ produces the solution $x \geq -2$. This means that -2 and any number greater than -2 produces a true statement. Solution sets for linear inequalities will often be displayed using a number line. If a value is included in the set ($\geq$ or $\leq$), a shaded dot is placed on that value and an arrow extending in the direction of the solutions. For a variable > or $\geq$ a number, the arrow will point right on a number line, the direction where the numbers increase. If a variable is < or $\leq$ a number, the arrow will point left on a number line, which is the direction where the numbers decrease. If the value is not included in the set (> or <), an open (unshaded) circle on that value is used with an arrow in the appropriate direction.

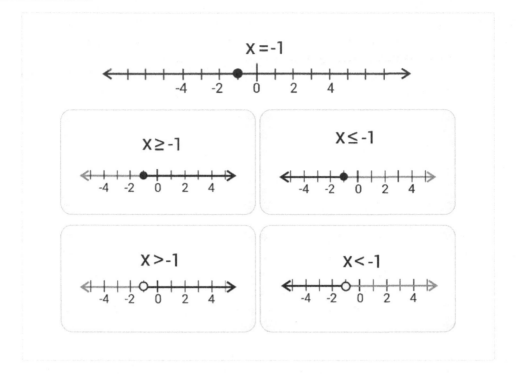

Similar to linear equations, a linear inequality may have a solution set consisting of all real numbers or can contain no solution. When solved algebraically, a linear inequality in which the variable cancels out and results in a true statement (ex. $7 \geq 2$) has a solution set of all real numbers. A linear inequality in which the variable cancels out and results in a false statement (ex. $7 \leq 2$) has no solution.

Linear Inequalities in One Variable

Linear inequalities and linear equations are both comparisons of two algebraic expressions. However, unlike equations in which the expressions are equal to each other, **linear inequalities** compare expressions that are unequal. Linear equations typically have one value for the variable that makes the statement true. Linear inequalities generally have an infinite number of values that make the statement true.

Linear inequalities are a concise mathematical way to express the relationship between unequal values. More specifically, they describe in what way the values are unequal. A value could be greater than ($>$); less than ($<$); greater than or equal to ($\geq$); or less than or equal to ($\leq$) another value. The statement "five times a number added to forty is more than sixty-five" can be expressed as $5x + 40 > 65$. Common words and phrases that express inequalities are:

Symbol	Phrase
$<$	is under, is below, smaller than, beneath
$>$	is above, is over, bigger than, exceeds
$\leq$	no more than, at most, maximum
$\geq$	no less than, at least, minimum

Solving Linear Inequalities

When solving a linear inequality, the solution is the set of all numbers that makes the statement true. The inequality $x + 2 \geq 6$ has a solution set of 4 and every number greater than 4 (4.0001, 5, 12, 107,

etc.). Adding 2 to 4 or any number greater than 4 would result in a value that is greater than or equal to 6. Therefore, $x \geq 4$ would be the solution set.

Solution sets for linear inequalities often will be displayed using a number line. If a value is included in the set ($\geq$ or $\leq$), there is a shaded dot placed on that value and an arrow extending in the direction of the solutions. For a variable $>$ or $\geq$ a number, the arrow would point right on the number line (the direction where the numbers increase); and if a variable is $<$ or $\leq$ a number, the arrow would point left (where the numbers decrease). If the value is not included in the set ($>$ or $<$), an open circle on that value would be used with an arrow in the appropriate direction.

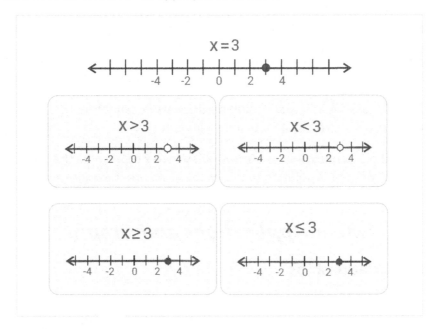

Students may be asked to write a linear inequality given a graph of its solution set. To do so, they should identify whether the value is included (shaded dot or open circle) and the direction in which the arrow is pointing.

In order to algebraically solve a linear inequality, the same steps should be followed as in solving a linear equation. The inequality symbol stays the same for all operations EXCEPT when dividing by a negative number. If dividing by a negative number while solving an inequality, the relationship reverses (the sign flips). Dividing by a positive does not change the relationship, so the sign stays the same. In other words, $>$ switches to $<$ and vice versa. An example is shown below.

Solve $-2(x + 4) \leq 22$

Distribute: $-2x - 8 \leq 22$

Add 8 to both sides: $-2x \leq 30$

Divide both sides by -2: $x \geq 15$

Systems of Two Linear Equations in Two Variables

A system of two linear equations in two variables is a set of equations that use the same variables, usually x and y. Here's a sample problem:

An Internet provider charges an installation fee and a monthly charge. It advertises that two months of its offering costs $100 and six months costs $200. Find the monthly charge and the installation fee.

The two unknown quantities (variables) are the monthly charge and the installation fee. There are two different statements given relating the variables: two months added to the installation fee is $100; and six months added to the installation fee is $200. Using the variable x as the monthly charge and y as the installation fee, the statements can be written as the following: $2x + y = 100; 6x + y = 200$. These two equations taken together form a system modeling the given scenario.

Systems of Linear Inequalities in Two Variables

A system of linear inequalities consists of two linear inequalities making comparisons between two variables. Students may be given a scenario and asked to express it as a system of inequalities:

> A consumer study calls for at least 60 adult participants. It cannot use more than 25 men. Express these constraints as a system of inequalities.

This can be modeled by the system: $x + y \geq 60; x \leq 25$, where x represents the number of men and y represents the number of women. A solution to the system is an ordered pair that makes both inequalities true when substituting the values for x and y.

Linear Applications and Graphs

Linear Models and Relationships

Linear relationships describe the way two quantities change with respect to each other. The relationship is defined as linear because a line is produced if all the sets of corresponding values are graphed on a coordinate grid. When expressing the linear relationship as an equation, the equation is often written in the form $y = mx + b$ (slope-intercept form) where m and b are numerical values and x and y are variables (for example, $y = 5x + 10$). Given a linear equation and the value of either variable (x or y), the value of the other variable can be determined.

Suppose a teacher is grading a test containing 20 questions with 5 points given for each correct answer, adding a curve of 10 points to each test. This linear relationship can be expressed as the equation $y = 5x + 10$ where x represents the number of correct answers and y represents the test score. To determine the score of a test with a given number of correct answers, the number of correct answers is substituted into the equation for x and evaluated. For example, for 10 correct answers, 10 is substituted for x:

$$y = 5(10) + 10 \rightarrow y = 60$$

Therefore, 10 correct answers will result in a score of 60. The number of correct answers needed to obtain a certain score can also be determined. To determine the number of correct answers needed to score a 90, 90 is substituted for y in the equation (y represents the test score) and solved:

$$90 = 5x + 10 \rightarrow 80 = 5x \rightarrow 16 = x$$

Therefore, 16 correct answers are needed to score a 90.

Linear relationships may be represented by a table of 2 corresponding values. Certain tables may determine the relationship between the values and predict other corresponding sets. Consider the table below, which displays the money in a checking account that charges a monthly fee:

Month	0	1	2	3	4
Balance	$210	$195	$180	$165	$150

An examination of the values reveals that the account loses $15 every month (the month increases by one and the balance decreases by 15). This information can be used to predict future values. To determine what the value will be in month 6, the pattern can be continued, and it can be concluded that the balance will be $120. To determine which month the balance will be $0, $210 is divided by $15 (since the balance decreases $15 every month), resulting in month 14.

Similar to a table, a graph can display corresponding values of a linear relationship.

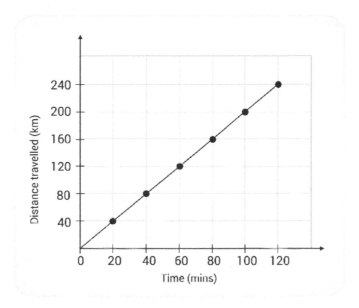

The graph above represents the relationship between distance traveled and time. To find the distance traveled in 80 minutes, the mark for 80 minutes is located at the bottom of the graph. By following this mark directly up on the graph, the corresponding point for 80 minutes is directly across from the 160-kilometer mark. This information indicates that the distance travelled in 80 minutes is 160 kilometers. To predict information not displayed on the graph, the way in which the variables change with respect to one another is determined. In this case, distance increases by 40 kilometers as time increases by 20 minutes. This information can be used to continue the data in the graph or convert the values to a table.

Identifying Variables for Linear Models

The first step to writing a linear model is to identify what the variables represent. A **variable** represents an unknown quantity, and in the case of a linear equation, a specific relationship exists between the two variables (usually x and y). Within a given scenario, the variables are the two quantities that are changing. The variable x is considered the independent variable and represents the inputs of a function. The variable y is considered the dependent variable and represents the outputs of a function. For example, if a scenario describes distance traveled and time traveled, distance would be represented by y and time represented by x. The distance traveled depends on the time spent traveling (time is independent). If a scenario describes the cost of a cab ride and the distance traveled, the cost would be represented by y and the distance represented by x. The cost of a cab ride depends on the distance traveled.

Identifying the Slope and Y-Intercept for Linear Models

The **slope of the graph of a line** represents the rate of change between the variables of an equation. In the context of a real-world scenario, the slope will tell the way in which the unknown quantities (variables) change with respect to each other. A scenario involving distance and time might state that someone is traveling at a rate of 45 miles per hour. The slope of the linear model would be 45. A scenario involving the cost of a cab ride and distance traveled might state that the person is charged $3 for each mile. The slope of the linear model would be 3.

The **y-intercept of a linear function** is the value of y when $x = 0$ (the point where the line intercepts the y-axis on the graph of the equation). It is sometimes helpful to think of this as a "starting point" for a linear function. Suppose for the scenario about the cab ride that the person is told that the cab company charges a flat fee of $5 plus $3 for each mile. Before traveling any distance ($x = 0$), the cost is $5. The y-intercept for the linear model would be 5.

Identifying Ordered Pairs for Linear Models

A linear equation with two variables can be written given a point (ordered pair) and the slope or given two points on a line. An ordered pair gives a set of corresponding values for the two variables (x and y). As an example, for a scenario involving distance and time, it is given that the person traveled 112.5 miles in 2 ½ hours. Knowing that x represents time and y represents distance, this information can be written as the ordered pair (2.5, 112.5).

Linear Functions Model a Linear Relationship Between Two Quantities

Linear relationships between two quantities can be expressed in two ways: function notation or as a linear equation with two variables. The relationship is referred to as linear because its graph is represented by a line. For a relationship to be linear, both variables must be raised to the first power only.

Function/Linear Equation Notation

A **relation** is a set of input and output values that can be written as ordered pairs. A function is a relation in which each input is paired with exactly one output. The domain of a function consists of all inputs, and the range consists of all outputs. Graphing the ordered pairs of a linear function produces a straight line. An example of a function would be $f(x) = 4x + 4$, read "f of x is equal to four times x plus four." In this example, the input would be x and the output would be f(x). Ordered pairs would be represented as $(x,$ f(x)$)$. To find the output for an input value of 3, 3 would be substituted for x into the function as follows: $f(3) = 4(3) + 4$, resulting in $f(3) = 16$. Therefore, the ordered pair $(3, f(3)) = (3, 16)$. Note f(x) is a function of x denoted by f. Functions of x could be named $g(x)$, read "g of x"; $p(x)$, read "p of x"; etc.

A linear function could also be written in the form of an equation with two variables. Typically, the variable x represents the inputs and the variable y represents the outputs. The variable x is considered the independent variable and y the dependent variable. The above function would be written as $y = 4x + 4$. Ordered pairs are written in the form (x, y).

Writing Linear Equations in Two Variables

When writing linear equations in two variables, the process depends on the information given. Questions will typically provide the slope of the line and its *y*-intercept, an ordered pair and the slope, or two ordered pairs.

Linear equations are commonly written in slope-intercept form, $y = mx + b$, where *m* represents the slope of the line and *b* represents the *y*-intercept. The slope is the rate of change between the variables, usually expressed as a whole number or fraction. The *y*-intercept is the value of *y* when *x* = 0 (the point where the line intercepts the *y*-axis on a graph). Given the slope and *y*-intercept of a line, the values are substituted for *m* and *b* into the equation. A line with a slope of $\frac{1}{2}$ and *y*-intercept of -2 would have an equation:

$$y = \frac{1}{2}x - 2$$

The point-slope form of a line, $y - y_1 = m(x - x_1)$, is used to write an equation when given an ordered pair (point on the equation's graph) for the function and its rate of change (slope of the line). The values for the slope, *m*, and the point (x_1, y_1) are substituted into the point-slope form to obtain the equation of the line. A line with a slope of 3 and an ordered pair (4, -2) would have an equation:

$$y - (-2) = 3(x - 4)$$

If a question specifies that the equation be written in slope-intercept form, the equation should be manipulated to isolate *y*:

Solve: $y - (-2) = 3(x - 4)$

Distribute: $y + 2 = 3x - 12$

Subtract 2 from both sides: $y = 3x - 14$

Given two ordered pairs for a function, (x_1, y_1) and (x_2, y_2), it is possible to determine the rate of change between the variables (slope of the line). To calculate the slope of the line, m, the values for the ordered pairs should be substituted into the formula:

$$m = \frac{y_2 - y_1}{x_2 - x_1}$$

The expression is substituted to obtain a whole number or fraction for the slope. Once the slope is calculated, the slope and either of the ordered pairs should be substituted into the point-slope form to obtain the equation of the line.

Graphing a System of Two Linear Equations in Two Variables

A solution for a system of equations is an ordered pair that makes both equations true. One method for solving a system of equations is to graph both lines on a coordinate plane. If the lines intersect, the point of intersection is the solution to the system. Every point on a line represents an ordered pair that makes its equation true. The ordered pair represented by this point of intersection lies on both lines and therefore makes both equations true. This ordered pair should be checked by substituting its values into both of the original equations of the system. Note that given a system of equations and an ordered pair, the ordered pair can be determined to be a solution or not by checking it in both equations.

If, when graphed, the lines representing the equations of a system do not intersect, then the two lines are parallel to each other or they are the same exact line. Parallel lines extend in the same direction without ever meeting. A system consisting of parallel lines has no solution. If the equations for a system represent the same exact line, then every point on the line is a solution to the system. In this case, there would be an infinite number of solutions. A system consisting of intersecting lines is referred to as independent; a system consisting of parallel lines is referred to as inconsistent; and a system consisting of coinciding lines is referred to as dependent.

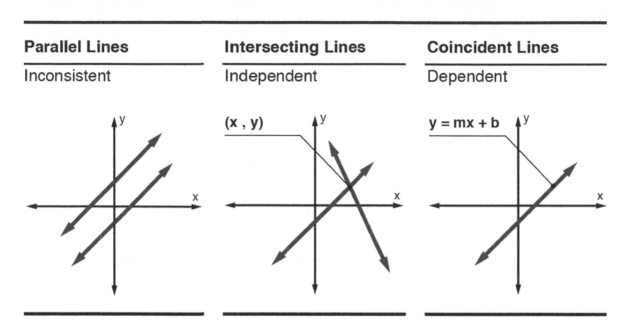

Parallel Lines	**Intersecting Lines**	**Coincident Lines**
Inconsistent	Independent	Dependent

Graphing Solution Sets for Linear Inequalities in Two Variables

A graph of the solution set for a linear inequality shows the ordered pairs that make the statement true. The graph consists of a boundary line dividing the coordinate plane and shading on one side of the boundary. The boundary line should be graphed just as a linear equation would be graphed. If the inequality symbol is $>$ or $<$, a dashed line can be used to indicate that the line is not part of the solution set. If the inequality symbol is $\geq$ or $\leq$, a solid line can be used to indicate that the boundary line is included in the solution set. An ordered pair (x, y) on either side of the line should be chosen to test in the inequality statement. If substituting the values for x and y results in a true statement ($15(3) + 25(2) > 90$), that ordered pair and all others on that side of the boundary line are part of the solution set. To indicate this, that region of the graph should be shaded. If substituting the ordered pair results in a false statement, the ordered pair and all others on that side are not part of the solution set.

Therefore, the other region of the graph contains the solutions and should be shaded.

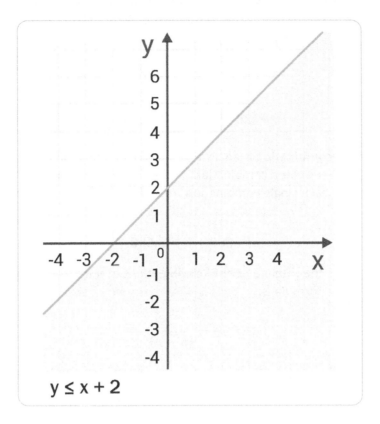

$y \leq x + 2$

A question may simply ask whether a given ordered pair is a solution to a given inequality. To determine this, the values should be substituted for the ordered pair into the inequality. If the result is a true statement, the ordered pair is a solution; if the result is a false statement, the ordered pair is not a solution.

Probability Sets

Chance Processes and Probability Models

Counting Techniques

There are many counting techniques that can help solve problems involving counting possibilities. For example, the **Addition Principle** states that if there are m choices from Group 1 and n choices from Group 2, then $n + m$ is the total number of choices possible from Groups 1 and 2. For this to be true, the groups can't have any choices in common. The **Multiplication Principle** states that if Process 1 can be completed n ways and Process 2 can be completed m ways, the total number of ways to complete both Process 1 and Process 2 is $n \times m$. For this rule to be used, both processes must be independent of each other. Counting techniques also involve permutations. A **permutation** is an arrangement of elements in a set for which order must be considered. For example, if three letters from the alphabet are chosen, ABC and BAC are two different permutations. The multiplication rule can be used to determine the total number of possibilities. If each letter can't be selected twice, the total number of possibilities is:

$$26 \times 25 \times 24 = 15,600$$

A formula can also be used to calculate this total. In general, the notation $P(n,r)$ represents the number of ways to arrange r objects from a set of n and, the formula is:

$$P(n,r) = \frac{n!}{(n-r)!}$$

In the previous example:

$$P(26,3) = \frac{26!}{23!} = 15,600$$

Contrasting permutations, a **combination** is an arrangement of elements in which order doesn't matter. In this case, ABC and BAC are the same combination. In the previous scenario, there are six permutations that represent each single combination. Therefore, the total number of possible combinations is:

$$15,600 \div 6 = 2,600$$

In general, $C(n,r)$ represents the total number of combinations of n items selected r at a time where order doesn't matter, and the formula is:

$$C(n,r) = \frac{n!}{(n-r)!\ r!}$$

Therefore, the following relationship exists between permutations and combinations:

$$C(n,r) = \frac{P(n,r)}{n!} = \frac{P(n,r)}{P(r,r)}$$

Fundamental Counting Principle

The **fundamental counting principle** states that if there are m possible ways for an event to occur, and n possible ways for a second event to occur, there are $m \cdot n$ possible ways for both events to occur. For example, there are two events that can occur after flipping a coin and six events that can occur after rolling a die, so there are $2 \times 6 = 12$ total possible event scenarios if both are done simultaneously. This principle can be used to find probabilities involving finite sample spaces and independent trials because it calculates the total number of possible outcomes. For this principle to work, the events must be independent of each other.

Independence and Conditional Probability

Sample Subsets

A sample can be broken up into subsets that are smaller parts of the whole. For example, consider a sample population of females. The sample can be divided into smaller subsets based on the characteristics of each female. There can be a group of females with brown hair and a group of females that wear glasses. There also can be a group of females that have brown hair *and* wear glasses. This "and" relates to the **intersection** of the two separate groups of brunettes and those with glasses. Every female in that intersection group has both characteristics. Similarly, there also can be a group of females that either have brown hair *or* wear glasses. The "or" relates to the union of the two separate groups of brunettes and glasses. Every female in this group has at least one of the characteristics. Finally, the group of females who do not wear glasses can be discussed. This "not" relates to the **complement** of the glass-wearing group. No one in the complement has glasses. **Venn diagrams** are useful in highlighting

these ideas. When discussing statistical experiments, this idea can also relate to events instead of characteristics.

Verifying Independent Events

Two events aren't always independent. For example, females with glasses and brown hair aren't independent characteristics. There definitely can be overlap because females with brown hair can wear glasses. Also, two events that exist at the same time don't have to have a relationship. For example, even if all females in a given sample are wearing glasses, the characteristics aren't related. In this case, the probability of a brunette wearing glasses is equal to the probability of a female being a brunette multiplied by the probability of a female wearing glasses. This mathematical test of $P(A \cap B) = P(A)P(B)$ verifies that two events are independent.

Conditional Probability

Conditional probability is the probability that event A will happen given that event B has already occurred. An example of this is calculating the probability that a person will eat dessert once they have eaten dinner. This is different than calculating the probability of a person just eating dessert. The formula for the conditional probability of event A occurring given B is $P(A|B) = \frac{P(A \text{ and } B)}{P(B)}$, and it's defined to be the probability of both A and B occurring divided by the probability of event B occurring. If A and B are independent, then the probability of both A and B occurring is equal to $P(A)P(B)$, so $P(A|B)$ reduces to just $P(A)$. This means that A and B have no relationship, and the probability of A occurring is the same as the conditional probability of A occurring given B. Similarly, $P(B|A) = \frac{P(B \text{ and } A)}{P(A)} = P(B)$ if A and B are independent.

Independent Versus Related Events

To summarize, conditional probability is the probability that an event occurs given that another event has happened. If the two events are related, the probability that the second event will occur changes if the other event has happened. However, if the two events aren't related and are therefore independent, the first event to occur won't impact the probability of the second event occurring.

Measuring Probabilities with Two-Way Frequency Tables

When measuring event probabilities, two-way frequency tables can be used to report the raw data and then used to calculate probabilities. If the frequency tables are translated into relative frequency tables, the probabilities presented in the table can be plugged directly into the formulas for conditional probabilities. By plugging in the correct frequencies, the data from the table can be used to determine if events are independent or dependent.

Differing Probabilities

The probability that event A occurs differs from the probability that event A occurs given B. When working within a given model, it's important to note the difference. $P(A|B)$ is determined using the formula $P(A|B) = \frac{P(A \text{ and } B)}{P(B)}$ and represents the total number of A's outcomes left that could occur after B occurs. $P(A)$ can be calculated without any regard for B. For example, the probability of a student finding a parking spot on a busy campus is different once class is in session.

The Addition Rule

The probability of event A or B occurring isn't equal to the sum of each individual probability. The probability that both events can occur at the same time must be subtracted from this total. This idea is shown in the **addition rule**:

$$P(A \text{ or } B) = P(A) + P(B) - P(A \text{ and } B)$$

The addition rule is another way to determine the probability of compound events that aren't mutually exclusive. If the events are mutually exclusive, the probability of both A and B occurring at the same time is 0.

Computing Probabilities

Simple and Compound Events

A **simple event** consists of only one outcome. The most popular simple event is flipping a coin, which results in either heads or tails. A **compound event** results in more than one outcome and consists of more than one simple event. An example of a compound event is flipping a coin while tossing a die. The result is either heads or tails on the coin and a number from one to six on the die. The probability of a simple event is calculated by dividing the number of possible outcomes by the total number of outcomes. Therefore, the probability of obtaining heads on a coin is $\frac{1}{2}$, and the probability of rolling a 6 on a die is $\frac{1}{6}$. The probability of compound events is calculated using the basic idea of the probability of simple events. If the two events are independent, the probability of one outcome is equal to the product of the probabilities of each simple event. For example, the probability of obtaining heads on a coin and rolling a 6 is equal to:

$$\frac{1}{2} \times \frac{1}{6} = \frac{1}{12}$$

The probability of either A or B occurring is equal to the sum of the probabilities minus the probability that both A and B will occur. Therefore, the probability of obtaining either heads on a coin or rolling a 6 on a die is:

$$\frac{1}{2} + \frac{1}{6} - \frac{1}{12} = \frac{7}{12}$$

The two events aren't mutually exclusive because they can happen at the same time. If two events are mutually exclusive, and the probability of both events occurring at the same time is zero, the probability of event A or B occurring equals the sum of both probabilities. An example of calculating the probability of two mutually exclusive events is determining the probability of pulling a king or a queen from a deck of cards. The two events cannot occur at the same time.

Uniform and Non-Uniform Probability Models

A **uniform probability model** is one where each outcome has an equal chance of occurring, such as the probabilities of rolling each side of a die. A **non-uniform probability model** is one where each outcome has an unequal chance of occurring. In a uniform probability model, the conditional probability formulas for $P(B|A)$ and $P(A|B)$ can be multiplied by their respective denominators to obtain two formulas for $P(A \text{ and } B)$. Therefore, the multiplication rule is derived as:

$$P(A \text{ and } B) = P(A)P(B|A) = P(B)P(A|B)$$

In a model, if the probability of either individual event is known and the corresponding conditional probability is known, the multiplication rule allows the probability of the joint occurrence of A and B to be calculated.

Descriptive Statistics

Basic Statistics

The field of statistics describes relationships between quantities that are related, but not necessarily in a deterministic manner. For example, a graduating student's salary will often be higher when the student graduates with a higher GPA, but this is not always the case. Likewise, people who smoke tobacco are more likely to develop lung cancer, but, in fact, it is possible for non-smokers to develop the disease as well. **Statistics** describes these kinds of situations, where the likelihood of some outcome depends on the starting data.

Descriptive statistics involves analyzing a collection of data to describe its broad properties such average (or mean), what percent of the data falls within a given range, and other such properties. An example of this would be taking all of the test scores from a given class and calculating the average test score. **Inferential statistics** attempts to use data about a subset of some population to make inferences about the rest of the population. An example of this would be taking a collection of students who received tutoring and comparing their results to a collection of students who did not receive tutoring, then using that comparison to try to predict whether the tutoring program in question is beneficial.

To be sure that inferences have a high probability of being true for the whole population, the subset that is analyzed needs to resemble a miniature version of the population as closely as possible. For this reason, statisticians like to choose random samples from the population to study, rather than picking a specific group of people based on some similarity. For example, studying the incomes of people who live in Portland does not tell anything useful about the incomes of people who live in Tallahassee.

A statistical question is answered by collecting data with variability. Data consists of facts and/or statistics (numbers), and variability refers to a tendency to shift or change. Data is a broad term, inclusive of things like height, favorite color, name, salary, temperature, gas mileage, and language. Questions requiring data as an answer are not necessarily statistical questions. If there is no variability in the data, then the question is not statistical in nature. Consider the following examples: what is Mary's favorite color? How much money does your mother make? What was the highest temperature last week? How many miles did your car get on its last tank of gas? How much taller than Bob is Ed?

None of the above are statistical questions because each case lacks variability in the data needed to answer the question. The questions on favorite color, salary, and gas mileage each require a single piece of data, whether a fact or statistic. Therefore, variability is absent. Although the temperature question requires multiple pieces of data (the high temperature for each day), a single, distinct number is the answer. The height question requires two pieces of data, Bob's height and Ed's height, but no difference in variability exists between those two values. Therefore, this is not a statistical question. Statistical questions typically require calculations with data.

Consider the following statistical questions:

How many miles per gallon of gas does the 2016 Honda Civic get? To answer this question, data must be collected. This data should include miles driven and gallons used. Different cars, different drivers, and

67

different driving conditions will produce different results. Therefore, variability exists in the data. To answer the question, the mean (average) value could be determined.

Are American men taller than German men? To answer this question, data must be collected. This data should include the heights of American men and the heights of German men. All American men are not the same height and all German men are not the same height. Some American men are taller than some German men and some German men are taller than some American men. Therefore, variability exists in the data. To answer the question, the median values for each group could be determined and compared.

The following are more examples of statistical questions: What proportion of 4th graders have a favorite color of blue? How much money do teachers make? Is it colder in Boston or Chicago?

An **experiment** is the method in which a hypothesis is tested using a trial-and-error process. A cause and the effect of that cause are measured, and the hypothesis is accepted or rejected. Experiments are usually completed in a controlled environment where the results of a control population are compared to the results of a test population. The groups are selected using a randomization process in which each group has a representative mix of the population being tested. Finally, an **observational study** is similar to an experiment. However, this design is used when there cannot be a designed control and test population because of circumstances (e.g., lack of funding or unrealistic expectations). Instead, existing control and test populations must be used, so this method has a lack of randomization.

Statistics involves making decisions and predictions about larger data sets based on smaller data sets. Basically, the information from one part or subset can help predict what happens in the entire data set or population at large. The entire process involves guessing, and the predictions and decisions may not be 100 percent correct all of the time; however, there is some truth to these predictions, and the decisions do have mathematical support. The smaller data set is called a **sample** and the larger data set (in which the decision is being made) is called a **population**. A **random sample** is used as the sample, which is an unbiased collection of data points that represents the population as well as it can. There are many methods of forming a random sample, and all adhere to the fact that every potential data point has a predetermined probability of being chosen.

Describing Distributions

Mean, Median, and Mode
The center of a set of data (statistical values) can be represented by its mean, median, or mode. These are sometimes referred to as measures of central tendency.

Mean
The first property that can be defined for this set of data is the **mean**. This is the same as the average. To find the mean, add up all the data points, then divide by the total number of data points. For example, suppose that in a class of 10 students, the scores on a test were 50, 60, 65, 65, 75, 80, 85, 85, 90, 100. Therefore, the average test score will be:

$$\frac{50 + 60 + 65 + 65 + 75 + 80 + 85 + 85 + 90 + 100}{10} = 75.5$$

The mean is a useful number if the distribution of data is normal (more on this later), which roughly means that the frequency of different outcomes has a single peak and is roughly equally distributed on both sides of that peak. However, it is less useful in some cases where the data might be split or where

there are some outliers. **Outliers** are data points that are far from the rest of the data. For example, suppose there are 10 executives and 90 employees at a company. The executives make $1000 per hour, and the employees make $10 per hour.

Therefore, the average pay rate will be:

$$\frac{\$1000 \times 10 + \$10 \times 90}{100} = \$109 \; per \; hour$$

In this case, this average is not very descriptive since it's not close to the actual pay of the executives *or* the employees.

Median

Another useful measurement is the **median**. In a data set, the median is the point in the middle. The middle refers to the point where half the data comes before it and half comes after, when the data is recorded in numerical order. For instance, these are the speeds of the fastball of a pitcher during the last inning that he pitched (in order from least to greatest):

$$90, 92, 93, 93, 95, 96, 97, 97, 97$$

There are nine total numbers, so the middle or *median* number is the 5th one, which is 95.

In cases where the number of data points is an even number, then the average of the two middle points is taken. In the previous example of test scores, the two middle points are 75 and 80. Since there is no single point, the average of these two scores needs to be found. The average is:

$$\frac{75 + 80}{2} = 77.5$$

The median is generally a good value to use if there are a few outliers in the data. It prevents those outliers from affecting the "middle" value as much as when using the mean.

Since an outlier is a data point that is far from most of the other data points in a data set, this means an outlier also is any point that is far from the median of the data set. The outliers can have a substantial effect on the mean of a data set, but they usually do not change the median or mode, or do not change them by a large quantity. For example, consider the data set (3, 5, 6, 6, 6, 8). This has a median of 6 and a mode of 6, with a mean of $\frac{34}{6} \approx 5.67$. Now, suppose a new data point of 1000 is added so that the data set is now (3, 5, 6, 6, 6, 8, 1000). The median and mode, which are both still 6, remain unchanged. However, the average is now $\frac{1034}{7}$, which is approximately 147.7. In this case, the median and mode will be better descriptions for most of the data points.

The reason for outliers in a given data set is a complicated problem. It is sometimes the result of an error by the experimenter, but often they are perfectly valid data points that must be taken into consideration.

Mode

One additional measure to define for *X* is the **mode**. This is the data point that appears most frequently. If two or more data points all tie for the most frequent appearance, then each of them is considered a mode. In the case of the test scores, where the numbers were 50, 60, 65, 65, 75, 80, 85, 85, 90, 100, there are two modes: 65 and 85.

Quartiles and Percentiles

The **first quartile** of a set of data X refers to the largest value from the first ¼ of the data points. In practice, there are sometimes slightly different definitions that can be used, such as the median of the first half of the data points (excluding the median itself if there are an odd number of data points). The term also has a slightly different use: when it is said that a data point lies *in the first quartile*, it means it is less than or equal to the median of the first half of the data points. Conversely, if it lies *at* the first quartile, then it is equal to the first quartile.

When it is said that a data point lies in the **second quartile**, it means it is between the first quartile and the median.

The **third quartile** refers to data that lies between ½ and ¾ of the way through the data set. Again, there are various methods for defining this precisely, but the simplest way is to include all of the data that lie between the median and the median of the top half of the data.

Data that lies in the **fourth quartile** refers to all of the data above the third quartile.

Percentiles may be defined in a similar manner to quartiles. Generally, this is defined in the following manner:

If a data point lies **in the n-th percentile**, this means it lies in the range of the first *n*% of the data.

If a data point lies **at the n-th percentile**, then it means that *n*% of the data lies below this data point.

Standard Deviation

Given a data set X consisting of data points $(x_1, x_2, x_3, \dots x_n)$, the **variance** of X is defined to be:

$$\frac{\sum_{i=1}^{n}(x_i - \bar{X})^2}{n}$$

This means that the variance of X is the average of the squares of the differences between each data point and the mean of X.

Given a data set X consisting of data points $(x_1, x_2, x_3, \dots x_n)$, the **standard deviation** of X is defined to be:

$$s_x = \sqrt{\frac{\sum_{i=1}^{n}(x_i - \bar{X})^2}{n}}$$

In other words, the standard deviation is the square root of the variance.

Both the variance and the standard deviation are measures of how much the data tend to be spread out. When the standard deviation is low, the data points are mostly clustered around the mean. When the standard deviation is high, this generally indicates that the data are quite spread out, or else that there are a few substantial outliers.

As a simple example, compute the standard deviation for the data set (1, 3, 3, 5). First, compute the mean, which will be:

$$\frac{1 + 3 + 3 + 5}{4} = \frac{12}{4} = 3$$

Now, find the variance of X with the formula:

$$\sum_{i=1}^{4} (x_i - \bar{X})^2 = (1 - 3)^2 + (3 - 3)^2 + (3 - 3)^2 + (5 - 3)^2$$

$$-2^2 + 0^2 + 0^2 + 2^2 = 8$$

Therefore, the variance is $\frac{8}{4} = 2$. Taking the square root, the standard deviation will be $\sqrt{2}$.

Note that the standard deviation only depends upon the mean, not upon the median or mode(s). Generally, if there are multiple modes that are far apart from one another, the standard deviation will be high. A high standard deviation does not always mean there are multiple modes, however.

Describing a Set of Data

A set of data can be described in terms of its center, spread, shape and any unusual features. The center of a data set can be measured by its mean, median, or mode. The spread of a data set refers to how far the data points are from the center (mean or median). A data set with data points clustered around the center will have a small spread. A data set covering a wide range will have a large spread.

When a data set is displayed as a graph like the one below, the shape indicates if a sample is normally distributed, symmetrical, or has measures of skewness. When graphed, a data set with a normal distribution will resemble a bell curve.

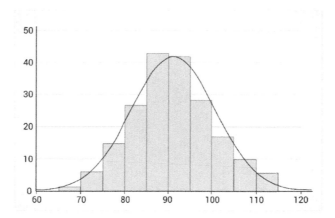

If the data set is symmetrical, each half of the graph when divided at the center is a mirror image of the other. If the graph has fewer data points to the right, the data is skewed right. If it has fewer data points to the left, the data is skewed left.

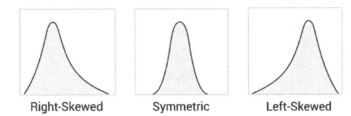

Right-Skewed   Symmetric   Left-Skewed

A description of a data set should include any unusual features such as gaps or outliers. A gap is a span within the range of the data set containing no data points. An outlier is a data point with a value either extremely large or extremely small when compared to the other values in the set.

The graphs above can be referred to as **unimodal** since they all have a single peak. In contrast, a bimodal graph has two peaks.

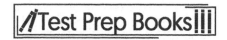

Geometry Concepts

Properties of Polygons and Circles

A **polygon** is a closed two-dimensional figure consisting of three or more sides. Polygons can be either convex or concave. A polygon that has interior angles all measuring less than 180° is **convex**. A **concave** polygon has one or more interior angles measuring greater than 180°. Examples are shown below.

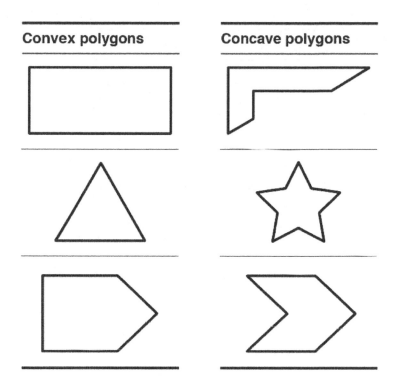

Polygons can be classified by the number of sides (also equal to the number of angles) they have. The following are the names of polygons with a given number of sides or angles:

# of sides	3	4	5	6	7	8	9	10
Name of polygon	Triangle	Quadrilateral	Pentagon	Hexagon	Septagon (or heptagon)	Octagon	Nonagon	Decagon

Equiangular polygons are polygons in which the measure of every interior angle is the same. The sides of equilateral polygons are always the same length. If a polygon is both equiangular and equilateral, the polygon is defined as a **regular polygon**. Examples are shown below.

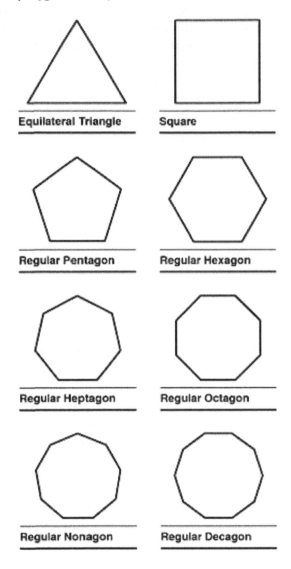

Equilateral Triangle

Square

Regular Pentagon

Regular Hexagon

Regular Heptagon

Regular Octagon

Regular Nonagon

Regular Decagon

Triangles can be further classified by their sides and angles. A triangle with its largest angle measuring 90° is a **right triangle**.

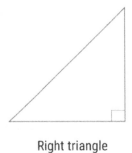

Right triangle

A triangle with the largest angle less than 90° is an **acute triangle**. A triangle with the largest angle greater than 90° is an **obtuse triangle**.

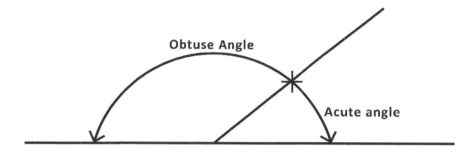

A triangle consisting of two equal sides and two equal angles is an **isosceles triangle**. A triangle with three equal sides and three equal angles is an **equilateral triangle**. A triangle with no equal sides or angles is a **scalene triangle**.

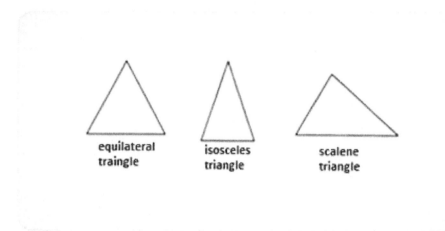

Quadrilaterals can be further classified according to their sides and angles. A quadrilateral with exactly one pair of parallel sides is called a **trapezoid**. A quadrilateral that shows both pairs of opposite sides parallel is a **parallelogram**. Parallelograms include rhombuses, rectangles, and squares. A **rhombus** has four equal sides. A **rectangle** has four equal angles (90° each). A **square** has four 90° angles and four equal sides. Therefore, a square is both a rhombus and a rectangle.

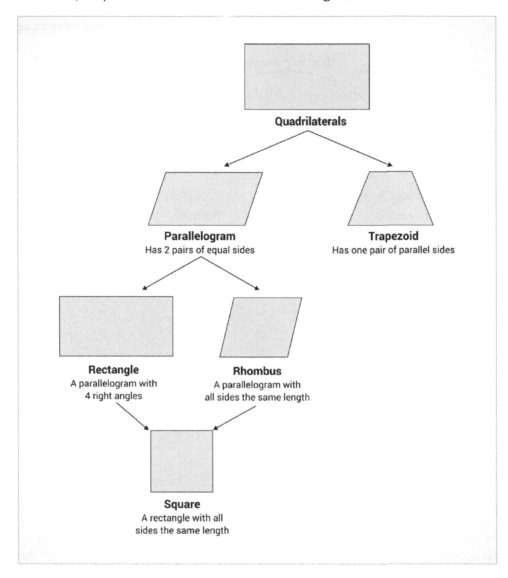

There are many key facts related to geometry that are applicable. The sum of the measures of the angles of a triangle are 180°, and for a quadrilateral, the sum is 360°. Rectangles and squares each have four right angles. A **right angle** has a measure of 90°.

Lines and Angles

In geometry, a **line** connects two points, has no thickness, and extends indefinitely in both directions beyond each point. If the length is finite, it's known as a **line segment** and has two **endpoints**. A **ray** is the straight portion of a line that has one endpoint and extends indefinitely in the other direction. An **angle** is formed when two rays begin at the same endpoint and extend indefinitely. The endpoint of an angle is called a **vertex**. **Adjacent angles** are two side-by-side angles formed from the same ray that have the same endpoint. Angles are measured in **degrees** or **radians**, which is a measure of **rotation**. A **full rotation** equals 360 degrees or 2π radians, which represents a circle. Half a rotation equals 180 degrees or π radians and represents a half-circle. Angle measurement is additive. When an angle is

broken into two non-overlapping angles, the total measure of the larger angle equals the sum of the two smaller angles. Lines are **coplanar** if they're located in the same plane. Two lines are **parallel** if they are coplanar, extend in the same direction, and never cross. If lines do cross, they're labeled as **intersecting lines** because they "intersect" at one point. If they intersect at more than one point, they're the same line. **Perpendicular lines** are coplanar lines that form a right angle at their point of intersection.

Two lines are parallel if they have the same slope and different intercept. Two lines are perpendicular if the product of their slope equals -1. Parallel lines never intersect unless they are the same line, and perpendicular lines intersect at a right angle. If two lines aren't parallel, they must intersect at one point. Determining equations of lines based on properties of parallel and perpendicular lines appears in word problems. To find an equation of a line, both the slope and a point the line goes through are necessary. Therefore, if an equation of a line is needed that's parallel to a given line and runs through a specified point, the slope of the given line and the point are plugged into the point-slope form of an equation of a line. Secondly, if an equation of a line is needed that's perpendicular to a given line running through a specified point, the negative reciprocal of the slope of the given line and the point are plugged into the point-slope form. Also, if the point of intersection of two lines is known, that point will be used to solve the set of equations. Therefore, to solve a system of equations, the point of intersection must be found. If a set of two equations with two unknown variables has no solution, the lines are parallel.

Classification of Angles

An **angle** consists of two rays that have a common endpoint. This common endpoint is called the **vertex** of the angle. The two rays can be called sides of the angle. The angle below has a vertex at point B and the sides consist of ray BA and ray BC. An angle can be named in three ways:

1. Using the vertex and a point from each side, with the vertex letter in the middle.
2. Using only the vertex. This can only be used if it is the only angle with that vertex.
3. Using a number that is written inside the angle.

The angle below can be written $\angle ABC$ (read angle ABC), $\angle CBA$, $\angle B$, or $\angle 1$.

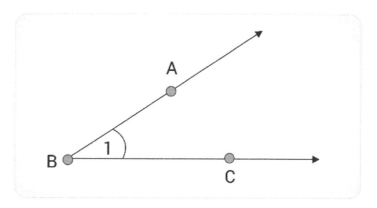

An angle divides a plane, or flat surface, into three parts: the angle itself, the interior (inside) of the angle, and the exterior (outside) of the angle. The figure below shows point *M* on the interior of the angle and point *N* on the exterior of the angle.

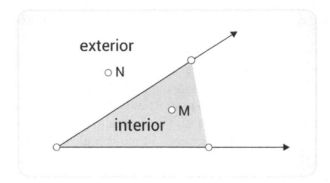

Angles can be measured in units called degrees, with the symbol °. The degree measure of an angle is between 0° and 180° and can be obtained by using a protractor.

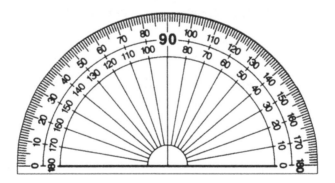

A straight angle (or simply a line) measures exactly 180°. A right angle's sides meet at the vertex to create a square corner. A right angle measures exactly 90° and is typically indicated by a box drawn in the interior of the angle. An acute angle has an interior that is narrower than a right angle. The measure of an acute angle is any value less than 90° and greater than 0°. For example, 89.9°, 47°, 12°, and 1°. An obtuse angle has an interior that is wider than a right angle. The measure of an obtuse angle is any value greater than 90° but less than 180°. For example, 90.1°, 110°, 150°, and 179.9°.

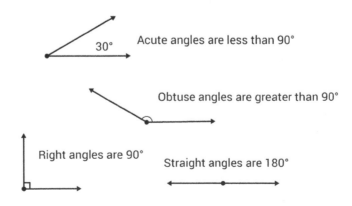

Perimeter and Area

Perimeter is the measurement of a distance around something or the sum of all sides of a polygon. Think of perimeter as the length of the boundary, like a fence. In contrast, **area** is the space occupied by a defined enclosure, like a field enclosed by a fence.

When thinking about perimeter, think about walking around the outside of something. When thinking about area, think about the amount of space or **surface area** something takes up.

Square

The perimeter of a square is measured by adding together all of the sides. Since a square has four equal sides, its perimeter can be calculated by multiplying the length of one side by 4. Thus, the formula is $P = 4 \times s$, where s equals one side. For example, the following square has side lengths of 5 meters:

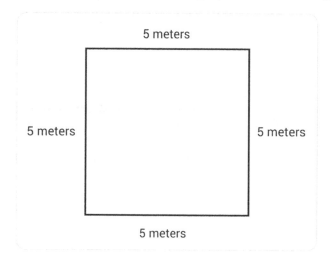

5 meters

5 meters   5 meters

5 meters

The perimeter is 20 meters because 4 times 5 is 20.

The area of a square is the length of a side squared. For example, if a side of a square is 7 centimeters, then the area is 49 square centimeters. The formula for this example is:

$$A = s^2 = 7^2 = 49 \text{ square centimeters}$$

An example is if the rectangle has a length of 6 inches and a width of 7 inches, then the area is 42 square inches:

$$A = lw = 6(7) = 42 \text{ square inches}$$

Rectangle

Like a square, a rectangle's perimeter is measured by adding together all of the sides. But as the sides are unequal, the formula is different. A rectangle has equal values for its lengths (long sides) and equal values for its widths (short sides), so the perimeter formula for a rectangle is:

$$P = l + l + w + w = 2l + 2w$$

l equals length
w equals width

The area is found by multiplying the length by the width, so the formula is $A = l \times w$.

For example, if the length of a rectangle is 10 inches and the width 8 inches, then the perimeter is 36 inches because:

$$P = 2l + 2w = 2(10) + 2(8) = 20 + 16 = 36 \text{ inches}$$

Triangle

A triangle's perimeter is measured by adding together the three sides, so the formula is $P = a + b + c$, where $a, b,$ and c are the values of the three sides. The area is the product of one-half the base and height so the formula is:

$$A = \frac{1}{2} \times b \times h$$

It can be simplified to:

$$A = \frac{bh}{2}$$

The base is the bottom of the triangle, and the height is the distance from the base to the peak. If a problem asks to calculate the area of a triangle, it will provide the base and height.

For example, if the base of the triangle is 2 feet and the height 4 feet, then the area is 4 square feet. The following equation shows the formula used to calculate the area of the triangle:

$$A = \frac{1}{2} bh = \frac{1}{2}(2)(4) = 4 \text{ square feet}$$

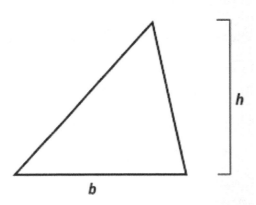

Circle

A circle's perimeter—also known as its circumference—is measured by multiplying the diameter by π.

Diameter is the straight line measured from a point on one side of the circle to a point directly across on the opposite side of the circle.

π is referred to as pi and is equal to 3.14 (with rounding).

So the formula is $\pi \times d$.

This is sometimes expressed by the formula $C = 2 \times \pi \times r$, where r is the radius of the circle. These formulas are equivalent, as the radius equals half of the diameter.

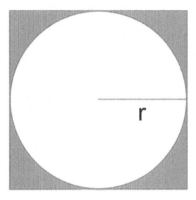

The area of a circle is calculated through the formula $A = \pi \times r^2$. The test will indicate either to leave the answer with π attached or to calculate to the nearest decimal place, which means multiplying by 3.14 for π.

The **arc of a circle** is the distance between two points on the circle. The length of the arc of a circle in terms of **degrees** is easily determined if the value of the central angle is known. The length of the arc is simply the value of the central angle. In this example, the length of the arc of the circle in degrees is 75°.

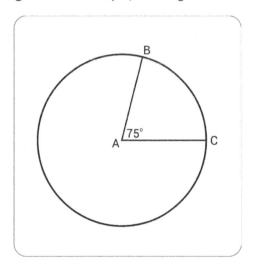

To determine the length of the arc of a circle in distance, the values for both the central angle and the radius must be known. This formula is:

$$\frac{central\ angle}{360°} = \frac{arc\ length}{2\pi r}$$

The equation is simplified by cross-multiplying to solve for the arc length.

In the following example, to solve for arc length, substitute the values of the central angle (75°) and the radius (10 inches) into the equation above.

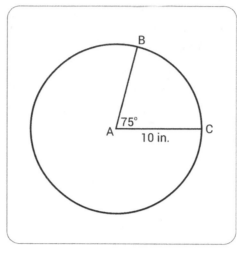

$$\frac{75°}{360°} = \frac{arc\ length}{2(3.14)(10in.)}$$

To solve the equation, first cross-multiply: 4710 = 360(arc length). Next, divide each side of the equation by 360. The result of the formula is that the arc length is 13.1 (rounded).

Irregular Shapes

The perimeter of an irregular polygon is found by adding the lengths of all of the sides. In cases where all of the sides are given, this will be very straightforward, as it will simply involve finding the sum of the provided lengths. Other times, a side length may be missing and must be determined before the perimeter can be calculated. Consider the example below:

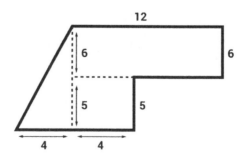

All of the side lengths are provided except for the angled side on the left. Test takers should notice that this is the hypotenuse of a right triangle. The other two sides of the triangle are provided (the base is 4 and the height is 6 + 5 = 11). The Pythagorean Theorem can be used to find the length of the hypotenuse, remembering that $a^2 + b^2 = c^2$.

Substituting the side values provided yields $(4)^2 + (11)^2 = c^2$.

Therefore, $c = \sqrt{16 + 121} = 11.7$

Finally, the perimeter can be found by adding this new side length with the other provided lengths to get the total length around the figure:

$$4+4+5+8+6+12+11.7=50.7$$

Although units are not provided in this figure, remember that reporting units with a measurement is important.

The area of an irregular polygon is found by decomposing, or breaking apart, the figure into smaller shapes. When the area of the smaller shapes is determined, these areas are added together to produce the total area of the area of the original figure. Consider the same example provided before:

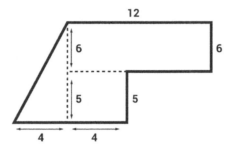

The irregular polygon is decomposed into two rectangles and a triangle. The area of the large rectangles ($A = l \times w \rightarrow A = 12 \times 6$) is 72 square units. The area of the small rectangle is 20 square units:

$$(A = 4 \times 5)$$

The area of the triangle ($A = \frac{1}{2} \times b \times h \rightarrow A = \frac{1}{2} \times 4 \times 11$) is 22 square units. The sum of the areas of these figures produces the total area of the original polygon:

$$A = 72 + 20 + 22 \rightarrow A = 114 \text{ square units}$$

Volume and Surface Area of Three-Dimensional Shapes

Geometry in three dimensions is similar to geometry in two dimensions. The main new feature is that three points now define a unique plane that passes through each of them. Three-dimensional objects can be made by putting together two-dimensional figures in different surfaces. Below, some of the possible three-dimensional figures will be provided, along with formulas for their volumes and surface areas.

Volume is the measurement of how much space an object occupies, like how much space is in the cube. Volume questions will ask how much of something is needed to completely fill the object. The most common surface area and volume questions deal with spheres, cubes, and rectangular prisms.

Surface area of a three-dimensional figure refers to the number of square units needed to cover the entire surface of the figure. This concept is similar to using wrapping paper to completely cover the outside of a box. For example, if a triangular pyramid has a surface area of 17 square inches (written $17in^2$), it will take 17 squares, each with sides one inch in length, to cover the entire surface of the pyramid. Surface area is also measured in square units.

A **rectangular prism** is a box whose sides are all rectangles meeting at 90° angles. Such a box has three dimensions: length, width, and height. If the length is x, the width is y, and the height is z, then the volume is given by $V = xyz$.

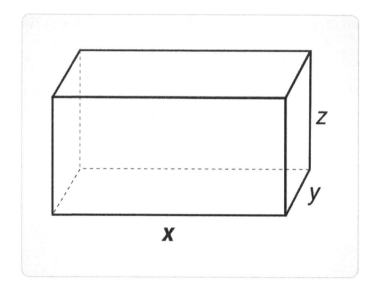

The surface area will be given by computing the surface area of each rectangle and adding them together. There is a total of six rectangles. Two of them have sides of length x and y, two have sides of length y and z, and two have sides of length x and z. Therefore, the total surface area will be given by:

$$SA = 2xy + 2yz + 2xz$$

A **cube** is a special type of rectangular solid in which its length, width, and height are the same. If this length is s, then the formula for the volume of a cube is $V = s \times s \times s$. The surface area of a cube is $SA = 6s^2$.

A **rectangular pyramid** is a figure with a rectangular base and four triangular sides that meet at a single vertex. If the rectangle has sides of length x and y, then the volume will be given by $V = \frac{1}{3}xyh$.

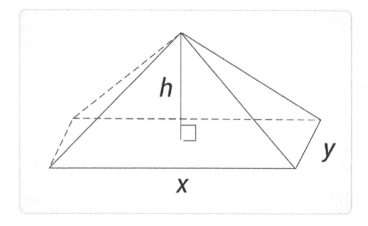

Many three-dimensional figures (solid figures) can be represented by nets consisting of rectangles and triangles. The surface area of such solids can be determined by adding the areas of each of its faces and

bases. Finding the surface area using this method requires calculating the areas of rectangles and triangles. To find the area (A) of a rectangle, the length (l) is multiplied by the width (w) → $A = l \times w$. The area of a rectangle with a length of 8cm and a width of 4cm is calculated: $A = (8cm) \times (4cm)$ → $A = 32cm^2$.

To calculate the area (A) of a triangle, the product of $\frac{1}{2}$, the base (b), and the height (h) is found:

$$A = \frac{1}{2} \times b \times h$$

Note that the height of a triangle is measured from the base to the vertex opposite of it forming a right angle with the base. The area of a triangle with a base of 11cm and a height of 6cm is calculated:

$$A = \frac{1}{2} \times (11cm) \times (6cm)$$

$$A = 33cm^2$$

Consider the following triangular prism, which is represented by a net consisting of two triangles and three rectangles.

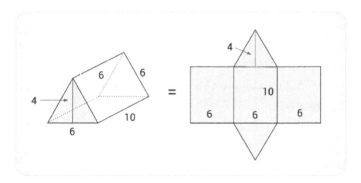

The surface area of the prism can be determined by adding the areas of each of its faces and bases. The surface area (SA) = area of triangle + area of triangle + area of rectangle + area of rectangle + area of rectangle.

$$SA = \left(\frac{1}{2} \times b \times h\right) + \left(\frac{1}{2} \times b \times h\right) + (l \times w) + (l \times w) + (l \times w)$$

$$SA = \left(\frac{1}{2} \times 6 \times 4\right) + \left(\frac{1}{2} \times 6 \times 4\right) + (6 \times 10) + (6 \times 10) + (6 \times 10)$$

$$SA = (12) + (12) + (60) + (60) + (60)$$

$$SA = 204 \; square \; units$$

A **sphere** is a set of points all of which are equidistant from some central point. It is like a circle, but in three dimensions. The volume of a sphere of radius r is given by:

$$V = \frac{4}{3}\pi r^3$$

The surface area is given by $A = 4\pi r^2$.

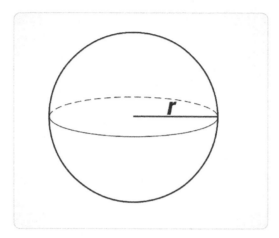

The volume of a **cylinder** is then found by adding a third dimension onto the circle. Volume of a cylinder is calculated by multiplying the area of the base (which is a circle) by the height of the cylinder. Doing so results in the equation $V = \pi r^2 h$. The volume of a **cone** is $\frac{1}{3}$ of the volume of a cylinder. Therefore, the formula for the volume of a cone is:

$$\frac{1}{3}\pi r^2 h$$

Plane Geometry

Algebraic equations can be used to describe geometric figures in the plane. The method for doing so is to use the **Cartesian coordinate plane**. The idea behind these Cartesian coordinates (named for mathematician and philosopher Descartes) is that from a specific point on the plane, known as the **origin**, one can specify any other point by saying *how far to the right or left* and *how far up or down*.

The plane is covered with a grid. The two directions, right to left and bottom to top, are called **axes** (singular **axis**). When working with x and y variables, the x variable corresponds to the right and left axis, and the y variable corresponds to the up and down axis.

Any point on the grid is found by specifying how far to travel from the center along the x-axis and how far to travel along the y-axis. The ordered pair can be written as (x, y). A positive x value means go to

the right on the *x*-axis, while a negative *x* value means to go to the left. A positive *y* value means to go up, while a negative value means to go down. Several points are shown as examples in the figure.

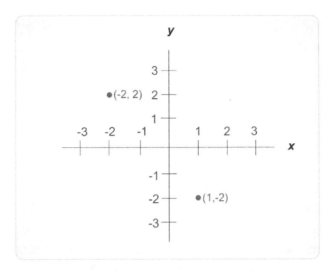

Cartesian Coordinate Plane

The coordinate plane can be divided into four **quadrants**. The upper-right part of the plane is called the first quadrant, where both *x* and *y* are positive. The second quadrant is the upper-left, where x is negative but y is positive. The third quadrant is the lower left, where both x and y are negative. Finally, the fourth quadrant is in the lower right, where x is positive but y is negative. These quadrants are often written with Roman numerals:

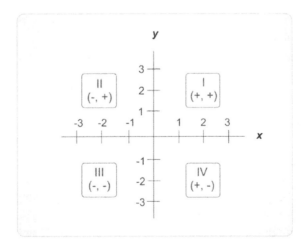

In addition to graphing individual points as shown above, the graph lines and curves in the plane can be graphed corresponding to equations. In general, if there is some equation involving x and y, then the graph of that equation consists of all the points (x, y) in the Cartesian coordinate plane, which satisfy this equation.

Given the equation $y = x + 2$, the point $(0, 2)$ is in the graph, since $2 = 0 + 2$ is a true equation. However, the point $(1, 4)$ will *not* be in the graph, because $4 = 1 + 2$ is false.

Proving Theorems with Coordinates

Many important formulas and equations exist in geometry that use coordinates. The distance between two points (x_1, y_1) and (x_2, y_2) is:

$$d = \sqrt{(x_2 - x_1)^2 + (y_2 - y_1)^2}$$

The slope of the line containing the same two points is $m = \frac{y_2 - y_1}{x_2 - x_1}$. Also, the midpoint of the line segment with endpoints (x_1, y_1) and (x_2, y_2) is:

$$M = \left(\frac{x_1 + x_2}{2}, \frac{y_1 + y_2}{2}\right)$$

The equations of a circle, parabola, ellipse, and hyperbola can also be used to prove theorems algebraically. Knowing when to use which formula or equation is extremely important, and knowing which formula applies to which property of a given geometric shape is an integral part of the process. In some cases, there are a number of ways to prove a theorem; however, only one way is required.

Formulas for Ratios

If a line segment with endpoints (x_1, y_1) and (x_2, y_2) is partitioned into two equal parts, the formula for midpoint is used. Recall this formula is $M = \left(\frac{x_1 + x_2}{2}, \frac{y_1 + y_2}{2}\right)$ and the ratio of line segments is 1:1. However, if the ratio needs to be anything other than 1:1, a different formula must be used. Consider a ratio that is $a : b$. This means the desired point that partitions the line segment is $\frac{a}{a+b}$ of the way from (x_1, y_1) to (x_2, y_2). The actual formula for the coordinate is:

$$\left(\frac{bx_1 + ax_2}{a + b}, \frac{by_1 + ay_2}{a + b}\right)$$

Computing Side Length, Perimeter, and Area

The side lengths of each shape can be found by plugging the endpoints into the distance formula:

$$d = \sqrt{(x_2 - x_1)^2 + (y_2 - y_1)^2}$$

between two ordered pairs (x_1, y_1) and (x_2, y_2). The distance formula is derived from the Pythagorean theorem. Once the side lengths are found, they can be added together to obtain the perimeter of the given polygon. Simplifications can be made for specific shapes such as squares and equilateral triangles. For example, one side length can be multiplied by 4 to obtain the perimeter of a square. Also, one side length can be multiplied by 3 to obtain the perimeter of an equilateral triangle. A similar technique can be used to calculate areas. For polygons, both side length and height can be found by using the same distance formula. Areas of triangles and quadrilaterals are straightforward through the use of $A = \frac{1}{2}bh$ or $A = bh$, depending on the shape. To find the area of other polygons, their shapes can be partitioned into rectangles and triangles. The areas of these simpler shapes can be calculated and then added together to find the total area of the polygon.

Transformations

Given a figure drawn on a plane, many changes can be made to that figure, including rotation, translation, and reflection. **Rotations** turn the figure about a point, **translations** slide the figure, and **reflections** flip the figure over a specified line. When performing these transformations, the original figure is called the **pre-image**, and the figure after transformation is called the **image**.

More specifically, **translation** means that all points in the figure are moved in the same direction by the same distance. In other words, the figure is slid in some fixed direction. Of course, while the entire figure is slid by the same distance, this does not change any of the measurements of the figures involved. The result will have the same distances and angles as the original figure.

In terms of Cartesian coordinates, a translation means a shift of each of the original points (x, y) by a fixed amount in the x and y directions, to become $(x + a, y + b)$.

Another procedure that can be performed is called **reflection**. To do this, a line in the plane is specified, called the **line of reflection**. Then, take each point and flip it over the line so that it is the same distance from the line but on the opposite side of it. This does not change any of the distances or angles involved, but it does reverse the order in which everything appears.

To reflect something over the x-axis, the points (x, y) are sent to $(x, -y)$. To reflect something over the y-axis, the points (x, y) are sent to the points $(-x, y)$. Flipping over other lines is not something easy to express in Cartesian coordinates. However, by drawing the figure and the line of reflection, the distance to the line and the original points can be used to find the reflected figure.

Example: Reflect this triangle with vertices (-1, 0), (2, 1), and (2, 0) over the y-axis. The pre-image is shown below.

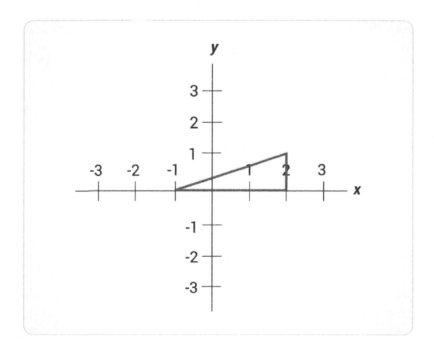

To do this, flip the *x* values of the points involved to the negatives of themselves, while keeping the *y* values the same. The image is shown here.

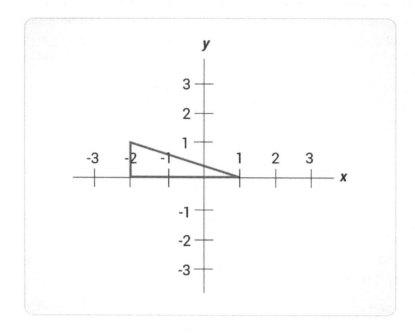

The new vertices will be (1, 0), (-2, 1), and (-2, 0).

Another procedure that does not change the distances and angles in a figure is **rotation**. In this procedure, pick a center point, then rotate every vertex along a circle around that point by the same angle. This procedure is also not easy to express in Cartesian coordinates, and this is not a requirement on this test. However, as with reflections, it's helpful to draw the figures and see what the result of the rotation would look like. This transformation can be performed using a compass and protractor.

Each one of these transformations can be performed on the coordinate plane without changes to the original dimensions or angles.

If two figures in the plane involve the same distances and angles, they are called **congruent figures**. In other words, two figures are congruent when they go from one form to another through reflection, rotation, and translation, or a combination of these.

Remember that rotation and translation will give back a new figure that is identical to the original figure, but reflection will give back a mirror image of it.

To recognize that a figure has undergone a rotation, check to see that the figure has not been changed into a mirror image, but that its orientation has changed (that is, whether the parts of the figure now form different angles with the *x* and *y* axes).

To recognize that a figure has undergone a translation, check to see that the figure has not been changed into a mirror image, and that the orientation remains the same.

To recognize that a figure has undergone a reflection, check to see that the new figure is a mirror image of the old figure.

Keep in mind that sometimes a combination of translations, reflections, and rotations may be performed on a figure.

Dilation

A **dilation** is a transformation that preserves angles, but not distances. This can be thought of as stretching or shrinking a figure. If a dilation makes figures larger, it is called an **enlargement**. If a dilation makes figures smaller, it is called a **reduction**. The easiest example is to dilate around the origin. In this case, multiply the x and y coordinates by a **scale factor**, k, sending points (x, y) to (kx, ky).

As an example, draw a dilation of the following triangle, whose vertices will be the points (-1, 0), (1, 0), and (1, 1).

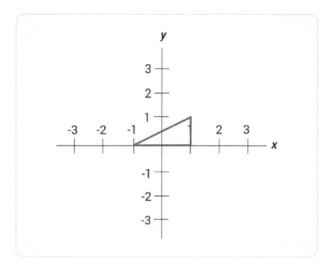

For this problem, dilate by a scale factor of 2, so the new vertices will be (-2, 0), (2, 0), and (2, 2).

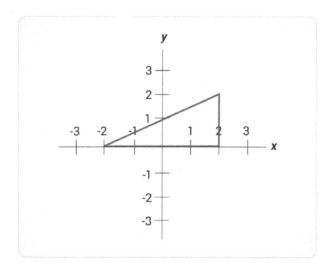

Note that after a dilation, the distances between the vertices of the figure will have changed, but the angles remain the same. The two figures that are obtained by dilation, along with possibly translation, rotation, and reflection, are all *similar* to one another. Another way to think of this is that similar figures

have the same number of vertices and edges, and their angles are all the same. Similar figures have the same basic shape but are different in size.

Symmetry

Using the types of transformations above, if an object can undergo these changes and not appear to have changed, then the figure is symmetrical. If an object can be split in half by a line and flipped over that line to lie directly on top of itself, it is said to have **line symmetry**. An example of both types of figures is seen below.

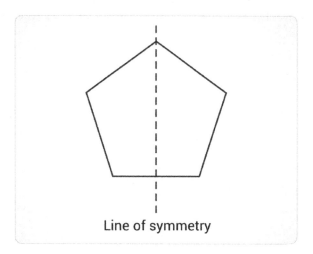

Line of symmetry

If an object can be rotated about its center to any degree smaller than 360, and it lies directly on top of itself, the object is said to have **rotational symmetry**. An example of this type of symmetry is shown below. The pentagon has an order of 5.

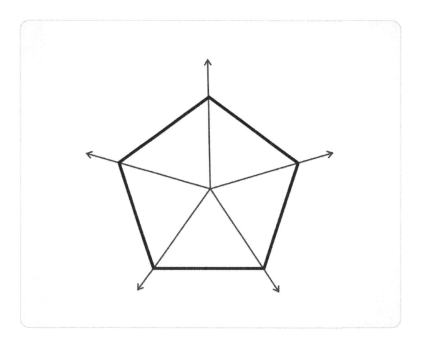

The rotational symmetry lines in the figure above can be used to find the angles formed at the center of the pentagon. Knowing that all of the angles together form a full circle, at 360 degrees, the figure can be split into 5 angles equally. By dividing the 360° by 5, each angle is 72°.

Given the length of one side of the figure, the perimeter of the pentagon can also be found using rotational symmetry. If one side length was 3 cm, that side length can be rotated onto each other side length four times. This would give a total of 5 side lengths equal to 3 cm. To find the perimeter, or distance around the figure, multiply 3 by 5. The perimeter of the figure would be 15 cm.

If a line cannot be drawn anywhere on the object to flip the figure onto itself or rotated less than or equal to 180 degrees to lay on top of itself, the object is asymmetrical. Examples of these types of figures are shown below.

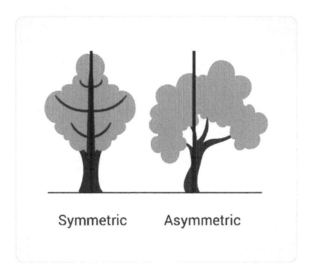

Symmetric   Asymmetric

No line of symmetry

Right Triangles: Pythagorean Theorem and Trigonometric Ratio

The value of a missing side of a right triangle may be determined two ways. The first way is to apply the Pythagorean Theorem, and the second way is to apply Trigonometric Ratios. The Pythagorean Theorem states that for every right triangle, the square of the length of the hypotenuse is equal to the sum of the squares of the lengths of the remaining two sides. The hypotenuse is the longest side of a right triangle and is also the side opposite the right angle.

According to the diagram, $a^2 + b^2 = c^2$, where c represents the hypotenuse, and a and b represent the lengths of the remaining two sides of the right triangle.

The Pythagorean Theorem may be applied a multitude of ways. For example, a person wishes to build a garden in the shape of a rectangle, having the dimensions of 5 feet by 8 feet. The garden's design includes a diagonal board to separate various types of plants. The Pythagorean Theorem can be used to determine the length of the diagonal board.

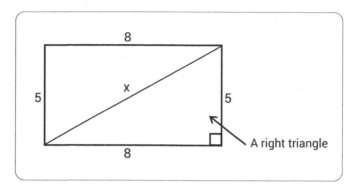

$$a^2 + b^2 = c^2$$

$$5^2 + 8^2 = c^2$$

$$25 + 64 = c^2$$

$$c = \sqrt{89}$$

$$c = 9.43$$

To solve for unknown sides of a right triangle using trigonometric ratios, the sine, cosine, and tangent are required.

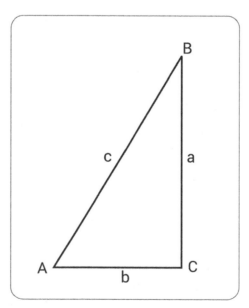

In the image above, angles are denoted by capital letters, and sides are denoted by lowercase letters. When examining angle *A*, *b* is the adjacent side, *a* is the opposite side, and *c* is the hypotenuse side. The various ratios of the lengths of the sides of the right triangle are used to find the sine, cosine, and tangent of angle *A*.

Thus, $\sin(A) = \frac{opposite}{hypotenuse}$, $\cos(A) = \frac{adjacent}{hypotenuse}$, and $\tan(A) = \frac{opposite}{adjacent}$. After substituting variables for the sides of the right triangle, $\sin(A) = \frac{a}{c}$, $\cos(A) = \frac{b}{c}$, and $\tan(A) = \frac{a}{b}$.

As a real-world example, the height of a tree can be discovered by using the information above. Surveying equipment can determine the tree's angle of inclination is 55.3 degrees, and the distance from the tree is 10 feet.

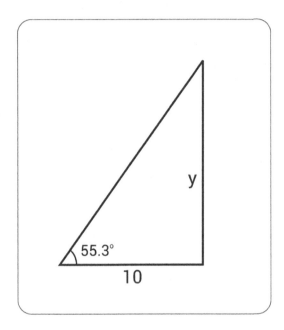

To find the height of the tree, substitute the known values into the trigonometric ratio of tangent:

$$\tan(55.3) = \frac{y}{10}$$

$$10 \times \tan(55.3) = y$$

$$10 \times 1.44418 = y$$

$$y = 14.4418$$

Similarity, Congruence, and Triangles

Triangles are similar if they have the same shape, the same angle measurements, and their sides are proportional to one another. Triangles are congruent if the angles of the triangles are equal in measurement and the sides of the triangles are equal in measurement.

There are five ways to show that a triangle is congruent.

- SSS (Side-Side-Side Postulate): When all three corresponding sides are equal in length, then the two triangles are congruent.

- SAS (Side-Angle-Side Postulate): If a pair of corresponding sides and the angle in between those two sides are equal, then the two triangles are congruent.

- ASA (Angle-Side-Angle Postulate): If a pair of corresponding angles are equal and the side within those angles are equal, then the two triangles are equal.

- AAS (Angle-Angle-Side Postulate): When a pair of corresponding angles for two triangles and a non-included side are equal, then the two triangles are congruent.

- HL (Hypotenuse-Leg Theorem): If two right triangles have the same hypotenuse length, and one of the other sides are also the same length, then the two triangles are congruent.

If two triangles are discovered to be similar or congruent, this information can assist in determining unknown parts of triangles, such as missing angles and sides.

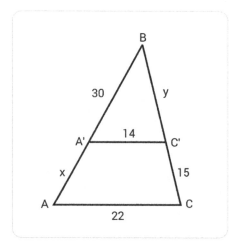

In the triangle shown above, AC and $A'C'$ are parallel lines. Therefore, BA is a transversal that intersects the two parallel lines. The corresponding angles $BA'C'$ and BAC are congruent. In a similar way, BC is also a transversal. Therefore, angle $BC'A'$ and BCA are congruent. If two triangles have two congruent angles, the triangles are similar. If the triangles are similar, their corresponding sides are proportional.

Therefore, the following equation is established:

$$\frac{30+x}{30} = \frac{22}{14} = \frac{y+15}{y}$$

$$\frac{30+x}{30} = \frac{22}{14}$$

$$x = 17.1$$

$$\frac{22}{14} = \frac{y+15}{y}$$

$$y = 26.25$$

The example below involves the question of congruent triangles. The first step is to examine whether the triangles are congruent. If the triangles are congruent, then the measure of a missing angle can be found.

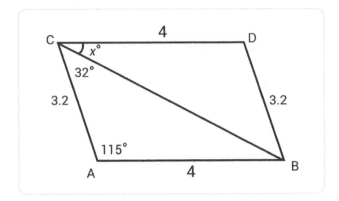

The above diagram provides values for angle measurements and side lengths in triangles *CAB* and *CDB*. Note that side *CA* is 3.2 and side *DB* is 3.2. Side *CD* is 4 and side *AB* is 4. Furthermore, line *CB* is congruent to itself by the reflexive property. Therefore, the two triangles are congruent by SSS (Side-Side-Side). Because the two triangles are congruent, all of the corresponding parts of the triangles are also congruent. Therefore, angle *x* is congruent to the inside of the angle for which a measurement is not provided in Triangle *CAB*. Thus, $115° + 32° = 147°$. A triangle measures 180°, therefore $180° - 147° = 33°$. *Angle x* = 33°, because the two triangles are reversed.

Complementary Angle Theorem

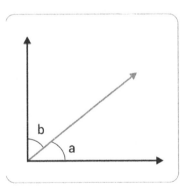

Two angles are complementary if the sum of the two angles equals 90°.

In the above diagram $Angle\ a + Angle\ b = 90°$. Therefore, the two angles are complementary. Certain trigonometric rules are also associated with complementary angles.

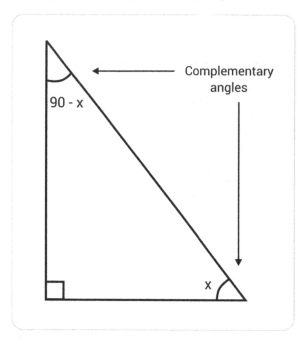

In the diagram above of a right triangle, if Angle A and Angle C are determined to be complementary angles, then certain relationships can be stated between the trigonometry of those angles.

$$sin(90° - x) = cos\ x$$

$$cos(90° - x) = sin\ x$$

For example, the sine of 80 degrees equals the cosine of $(90° - 80°)$, which is the cos $(10°)$.

This is true because the sine of an angle in a right triangle is equal to the cosine of its complement. Sine is known as the conjunction of cosine, and cosine is known as the conjunction of sine.

Examples:

1. $cos5° = sin\ x°$?
2. $sin(90° - x) = ?\ sin(90° - x) =?$

For problem number 1, the student should remember that $sin(90° - x) = cos\ x$. Cos 5° would be the same as $sin(90 - 5)°$. Therefore, $cos5° = sin85°$.

For problem number 2, the student would use the same fact that $sin(90° - x)° = cos\ x$.

An *acute angle* is an angle that is less than 90°. If Angle *A* and Angle *B* are acute angles of a right triangle, then $sinA = cosB$. Therefore, the sine of any acute angle in a right triangle is equal to the cosine of its complement, and the cosine of any acute angle is equal to the sine of its complement.

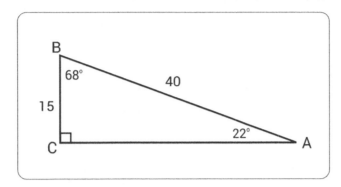

The example above is a right triangle. If only the value of angle *BAC* (which is 22°) was provided, the student would be able to figure out the value for angle *CBA* (68°) by knowing that a triangle is made up of 180°(180° − 90° − 22° = 68°). From the information given about acute angles on the previous page, the following statement is true:

Sine (angle *BAC*) = $\frac{15}{40}$, which is equivalent to the Cos (angle *CBA*) = $\frac{15}{40}$

Circles on the Coordinate Plane

If a circle is placed on the coordinate plane with the center of the circle at the origin (0,0), then point (*x*, *y*) is a point on the circle. Furthermore, the line extending from the center to point (*x*, *y*) is the radius, or *r*. By applying the Pythagorean Theorem $(a^2 + b^2 = c^2)$, it can be stated that $x^2 + y^2 = r^2$. However, the center of the circle does not always need to be on the origin of the coordinate plane.

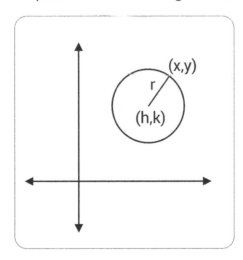

In the diagram above, the center of the circle is noted by (*h*, *k*). By applying the distance formula, the equation becomes: $= \sqrt{(x - h)^2 + (y - k)^2}$. When squaring both sides of the equation, the result is the standard form of a circle with the center (*h*, *k*) and radius *r*. Namely, $r^2 = (x - h)^2 + (y - k)^2$, where *r* = radius and center = (*h*, *k*). The following examples may be solved by using this information:

Example: Graph the equation $-x^2 + y^2 = 25$

To graph this equation, first note that the center of the circle is (0, 0). The radius is the positive square root of 25 or 5.

Example: Find the equation for the circle below.

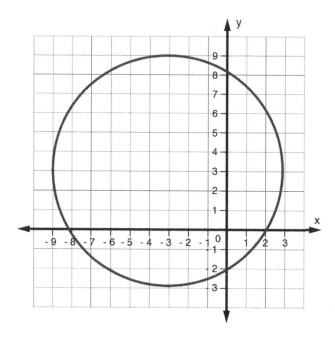

To find the equation for the circle, note that its center is not zero. Therefore, to find the circle's center, draw vertical and horizontal diameters to examine where they intersect. The center is located at point: (-3, 3). Next, count the number of spaces from the center to the outside of the circle. This number is 6. Therefore, 6 is the radius. Finally, plug in the numbers that are known into the standard equation for a circle:

$$36 = \left(x - (-3)\right)^2 + (y - 3)^2$$

or

$$36 = (x + 3)^2 + (y - 3)^2$$

It is possible to determine whether a point lies on a circle or not within the coordinate plane. For example, a circle has a center of (2, -5), and a radius of 6 centimeters. The first step is to apply the equation of a circle, which is $r^2 = (x - h)^2 + (y - k)^2$, where r = radius and the center = (h, k). Next, substitute the numbers for the center point and the number for the radius. This action simplifies the equation to $36 = (x - 2)^2 + (y + 5)^2$. Note that the radius of 6 was squared to get 36.

To prove that the point (2, -1) lies on the circle, apply the equation of the circle that was just used and input the values of (2, -1) for *x* and *y* in the equation.

$$36 = (x - 2)^2 + (y + 5)^2$$

$$36 = (2 - 2)^2 + (1 + 5)^2$$

$$36 = (0)^2 + (6)^2$$

$$36 = 36$$

Because the left side of the equation equals the right side of the equation, point (2, 1) lies on the given circle.

Practice Questions

1. The graph of which function has an x-intercept of -2?
   a. $y = 2x - 3$
   b. $y = 4x + 2$
   c. $y = x^2 + 5x + 6$
   d. $y = -\frac{1}{2} \times 2^x$

2. The table below displays the number of three-year-olds at Kids First Daycare who are potty-trained and those who still wear diapers.

	Potty-trained	Wear diapers	
Boys	26	22	48
Girls	34	18	52
	60	40	

What is the probability that a three-year-old girl chosen at random from the school is potty-trained?
   a. 52 percent
   b. 34 percent
   c. 65 percent
   d. 57 percent

3. What is the volume of a cylinder, in terms of π, with a radius of 5 inches and a height of 10 inches?
   a. $250\,\pi$ in³
   b. $50\,\pi$ in³
   c. $100\,\pi$ in³
   d. $200\,\pi$ in³

4. What is the function that forms an equivalent graph to $y = \cos(x)$?
   a. $y = \tan(x)$
   b. $y = \csc(x)$
   c. $y = \sin\left(x + \frac{\pi}{2}\right)$
   d. $y = \sin\left(x - \frac{\pi}{2}\right)$

5. A rectangle was formed out of pipe cleaner. Its length was $\frac{1}{2}$ feet and its width was $\frac{11}{2}$ inches. What is its area in square inches?
   a. $\frac{11}{4}$ inch²
   b. $\frac{11}{2}$ inch²
   c. 22 inch²
   d. 33 inch²

Answer Explanations

1. C: An x-intercept is the point where the graph crosses the x-axis. At this point, the value of y is 0. To determine if an equation has an x-intercept of -2, substitute -2 for x, and calculate the value of y. If the value of -2 for x corresponds with a y-value of 0, then the equation has an x-intercept of -2. The only answer choice that produces this result is Choice C:

$$0 = (-2)^2 + 5(-2) + 6$$

2. C: The conditional frequency of a girl being potty-trained is calculated by dividing the number of potty-trained girls by the total number of girls: $34 \div 52 = 0.65$. To determine the conditional probability, multiply the conditional frequency by 100: $0.65 \times 100 = 65\%$.

3. A: The volume of a cylinder is $\pi r^2 h$, and $\pi \times 5^2 \times 10$ is $250\,\pi\ in^3$. Choice B is not the correct answer because that is $5^2 \times 2\pi$. Choice C is not the correct answer since that is $5in \times 10\pi$. Choice D is not the correct answer because that is $10^2 \times 2in$.

4. C: Graphing the function $y = \cos(x)$ shows that the curve starts at $(0, 1)$, has an amplitude of 2, and a period of 2π. This same curve can be constructed using the sine graph, by shifting the graph to the left $\frac{\pi}{2}$ units. This equation is in the form:

$$y = \sin(x + \frac{\pi}{2})$$

5. D: Area = length x width. The answer must be in square inches, so all values must be converted to inches. $\frac{1}{2}$ ft is equal to 6 inches. Therefore, the area of the rectangle is equal to $6 \times \frac{11}{2} = \frac{66}{2} = 33$ square inches.

Advanced Algebra and Functions

Linear Equations

Solving Linear Equations

When asked to solve a linear equation, it requires determining a numerical value for the unknown variable. Given a linear equation involving addition, subtraction, multiplication, and division, isolation of the variable is done by working backward. Addition and subtraction are inverse operations, as are multiplication and division; therefore, they can be used to cancel each other out.

The first steps to solving linear equations are to distribute if necessary and combine any like terms that are on the same side of the equation. Sides of an equation are separated by an $=$ sign. Next, the equation should be manipulated to get the variable on one side. Whatever is done to one side of an equation, must be done to the other side to remain equal. Then, the variable should be isolated by using inverse operations to undo the order of operations backward. Undo addition and subtraction, then undo multiplication and division. For example:

Solve $4(t - 2) + 2t - 4 = 2(9 - 2t)$

Distribute: $4t - 8 + 2t - 4 = 18 - 4t$

Combine like terms: $6t - 12 = 18 - 4t$

Add 4t to each side to move the variable: $10t - 12 = 18$

Add 12 to each side to isolate the variable: $10t = 30$

Divide each side by 10 to isolate the variable: $t = 3$

The answer can be checked by substituting the value for the variable into the original equation and ensuring both sides calculate to be equal.

Algebraically Solving Linear Equations (or Inequalities) in One Variable

Linear equations in one variable and linear inequalities in one variable can be solved following similar processes. Although they typically have one solution, a linear equation can have no solution or can have a solution set of all real numbers. Solution sets for linear inequalities typically consist of an infinite number of values either greater or less than a given value (where the given value may or may not be included in the set). However, a linear inequality can have no solution or can have a solution set consisting of all real numbers.

Linear Equations in One Variable – Special Cases

Solving a linear equation produces a value for the variable that makes the algebraic statement true. If there is no value for the variable that would make the statement true, there is no solution to the equation. Here's a sample equation: $x + 3 = x - 1$. There is no value for x in which adding 3 to the value would produce the same result as subtracting 1 from that value. Conversely, if any value for the variable would make a true statement, the equation has an infinite number of solutions. Here's another

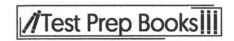

sample equation: $3x + 6 = 3(x + 2)$. Any real number substituted for x would result in a true statement (both sides of the equation are equal).

By manipulating equations similar to the two above, the variable of the equation will cancel out completely. If the constants that are left express a true statement (ex., $6 = 6$), then all real numbers are solutions to the equation. If the constants left express a false statement (ex., $3 = -1$), then there is no solution to the equation.

A question on this material may present a linear equation with an unknown value for either a constant or a coefficient of the variable and ask to determine the value that produces an equation with no solution or infinite solutions. For example:

$3x + 7 = 3x + 10 + n$; Find the value of n that would create an equation with an infinite number of solutions for the variable x.

To solve this problem, the equation should be manipulated so the variable x will cancel. To do this, $3x$ should be subtracted from both sides, which would leave $7 = 10 + n$. By subtracting 10 on both sides, it is determined that $n = -3$. Therefore, a value of -3 for n would result in an equation with a solution set of all real numbers.

If the same problem asked for the equation to have no solution, the value of n would be all real numbers except -3.

Linear Inequalities in One Variable – Special Cases

A linear inequality can have a solution set consisting of all real numbers or can contain no solution. When solved algebraically, a linear inequality in which the variable cancels out and results in a true statement (ex., $7 \geq 2$) has a solution set of all real numbers. A linear inequality in which the variable cancels out and results in a false statement (ex., $7 \leq 2$) has no solution.

Compound Inequalities

A compound inequality is a pair of inequalities joined by *and* or *or*. Given a compound inequality, to determine its solution set, both inequalities should be solved for the given variable. The solution set for a compound inequality containing *and* consists of all the values for the variable that make both inequalities true. If solving the compound inequality results in $x > -9$ and $x \leq 6$, the solution set would consist of all values between -2 and 3, including 3. This may also be written as follows: $-9 < x \leq 6$. Due to the graphs of their solution sets (shown below), compound inequalities such as these are referred to as conjunctions.

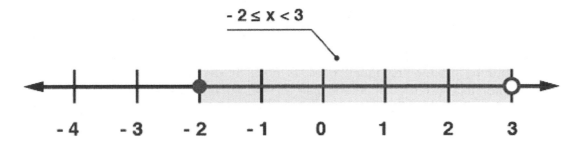

If there are no values that would make both inequalities of a compound inequality containing *and* true, then there is no solution. An example would be $x > 2$ and $x \leq 0$.

The solution set for a compound inequality containing *or* consists of all the values for the variable that make at least one of the inequalities true. The solution set for the compound inequality $x \leq -2$ or $x > 1$ consists of all values less than 2, 6, and all values greater than 6. Due to the graphs of their solution sets (shown below), compound inequalities such as these are referred to as disjunctions.

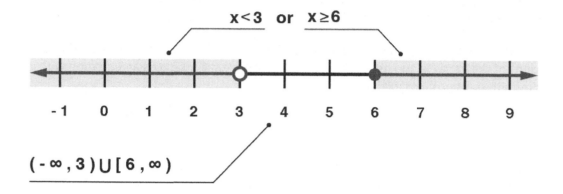

If the two inequalities for a compound inequality containing *or* "overlap," then the solution set contains all real numbers. An example would be $x > 2$ or $x < 7$. Any number would make at least one of these true.

Solving Systems of Two Linear Equations in Two Variables

A system of two linear equations in two variables is a set of equations that use the same variables (typically *x* and *y*). A solution to the system is an ordered pair that makes both equations true. One method for solving a system is by graphing. This method, however, is not always practical. Students may not have graph paper; or the solution may not consist of integers, making it difficult to identify the exact point of intersection on a graph. There are two methods for solving systems of equations algebraically: substitution and elimination. The method used will depend on the characteristics of the equations in the system.

Solving Systems of Equations with the Substitution Method

If one of the equations in a system has an isolated variable (*x*= or *y*=) or a variable that can be easily isolated, the substitution method can be used. Here's a sample system: $x + 3y = 7; 2x - 4y = 24$. The first equation can easily be solved for *x*. By subtracting $3y$ on both sides, the resulting equation is $x = 7 - 3y$. When one equation is solved for a variable, the expression that it is equal can be substituted into the other equation. For this example, $(7 - 3y)$ would be substituted for *x* into the second equation as follows: $2(7 - 3y) + 4y = 24$. Solving this equation results in $y = -5$. Once the value for one variable is known, this value should be substituted into either of the original equations to determine the value of the other variable. For the example, -5 would be substituted for *y* in either of the original equations. Substituting into the first equation results in $x + 3(-5) = 7$, and solving this equation yields $x = 22$. The solution to a system is an ordered pair, so the solution to the example is written as (22, 7). The solution can be checked by substituting it into both equations of the system to ensure it results in two true statements.

Solving Systems of Equations with the Elimination Method

The elimination method for solving a system of equations involves canceling out (or eliminating) one of the variables. This method is typically used when both equations of a system are written in standard

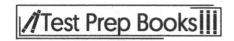

form $(Ax + By = C)$. An example is $2x + 3y = 12; 5x - y = 13$. To perform the elimination method, the equations in the system should be arranged vertically to be added together and then one or both of the equations should be multiplied so that one variable will be eliminated when the two are added. Opposites will cancel each other when added together. For example, 8x and -8x will cancel each other when added. For the example above, writing the system vertically helps identify that the bottom equation should be multiplied by 3 to eliminate the variable y.

$$2x + 3y = 12 \quad \rightarrow \quad 2x + 3y = 12$$

$$3(5x - y = 13) \quad \rightarrow \quad 15x - 3y = 39$$

Adding the two equations together vertically results in $17x = 51$. Solving yields $x = 3$. Once the value for one variable is known, it can be substituted into either of the original equations to determine the value of the other variable. Once this is obtained, the solution can be written as an ordered pair (x, y) and checked in both equations of the system. In this example, the solution is (3, 2).

Systems of Equations with No Solution or an Infinite Number of Solutions

A system of equations can have one solution, no solution, or an infinite number of solutions. If, while solving a system algebraically, both variables cancel out, then the system has either no solution or has an infinite number of solutions. If the remaining constants result in a true statement (ex., $7 = 7$), then there is an infinite number of solutions. This would indicate coinciding lines. If the remaining constants result in a false statement, then there is no solution to the system. This would indicate parallel lines.

Linear Applications and Graphs

Linear Growth and Decay

As mentioned, a linear equation can be written in the form $y = mx + b$, where x represents the inputs, y represents the outputs, b represents the y-intercept for the graph, and m represents the slope of the line. The y-intercept is the value of y when $x = 0$ and can be thought of as the "starting point." The slope is the rate of change between the variables x and y. A positive slope represents growth, and a negative slope represents decay. Given a table of values for inputs (x) and outputs (y), a linear function would model the relationship if: x and y change at a constant rate per unit interval—for every two inputs a given distance apart, the distance between their corresponding outputs is constant. Here are some sample ordered pairs:

x	0	1	2	3
y	-7	-4	-1	2

For every 1 unit increase in x, y increases by 3 units. Therefore, the change is constant and thus represents linear growth.

Given a scenario involving growth or decay, determining if there is a constant rate of change between inputs (x) and outputs (y) will identify if a linear model is appropriate. A scenario involving distance and time might state that someone is traveling at a rate of 45 miles per hour. For every hour traveled (input), the distance traveled (output) increases by 45 miles. This is a constant rate of change.

Graphing Linear Equations

The simplest equations to graph are the equations whose graphs are lines, called **linear equations**. Every linear equation can be rewritten algebraically so that it looks like:

$$Ax + By = C$$

First, the ratio of the change in the y coordinate to the change in the x coordinate is constant for any two distinct points on the line. In any pair of points on a line, two points, (x_1, y_1) and (x_2, y_2)—

where $x_1 \neq x_2$—the ratio $\frac{y_2 - y_1}{x_2 - x_1}$ will always be the same, even if another pair of points is used.

This ratio, $\frac{y_2 - y_1}{x_2 - x_1}$, is called the *slope* of the line and is often denoted with the letter m. If the slope is **positive**, then the line goes upward when moving to the right. If the slope is **negative**, then it moves downward when moving to the right. If the slope is 0, then the line is **horizontal**, and the y coordinate is constant along the entire line. For lines where the x coordinate is constant along the entire line, the slope is not defined, and these lines are called **vertical** lines.

The y coordinate of the point where the line touches the y-axis is called the **y-intercept** of the line. It is often denoted by the letter b, used in the form of the linear equation $y = mx + b$. The x coordinate of the point where the line touches the x-axis is called the **x-intercept**. It is also called the *zero* of the line.

Suppose two lines have slopes m_1 and m_2. If the slopes are equal, $m_1 = m_2$, then the lines are **parallel**. Parallel lines never meet one another. If $m_1 = -\frac{1}{m_2}$, then the lines are called **perpendicular** or **orthogonal**. Their slopes can also be called opposite reciprocals of each other.

There are several convenient ways to write down linear equations. The common forms are listed here:

- **Standard Form**: $Ax + By = C$, where the slope is given by $\frac{-A}{B}$, and the y-intercept is given by $\frac{C}{B}$.

- **Slope-Intercept Form**: $y = mx + b$, where the slope is m, and the y-intercept is b.

- **Point-Slope Form**: $y - y_1 = m(x - x_1)$, where m is the slope, and (x_1, y_1) is any point on the line.

- **Two-Point Form**: $\frac{y - y_1}{x - x_1} = \frac{y_2 - y_1}{x_2 - x_1}$, where (x_1, y_1), and (x_2, y_2) are any two distinct points on the line.

- **Intercept Form**: $\frac{x}{x_1} + \frac{y}{y_1} = 1$, where x_1 is the x-intercept, and y_1 is the y-intercept.

Depending upon the given information, different forms of the linear equation can be easier to write down than others. When given two points, the two-point form is easy to write down. If the slope and a single point is known, the point-slope form is easiest to start with. In general, which form to start with depends upon the given information.

Graphing a Line in Slope-Intercept Form

When an equation is written in slope-intercept form, $y = mx + b$, m represents the slope of the line and b represents the y-intercept. The y-intercept is the value of y when $x = 0$ and the point at which the graph of the line crosses the y-axis. The slope is the rate of change between the variables, expressed as a fraction. The fraction expresses the change in y compared to the change in x. If the slope is an integer, it should be written as a fraction with a denominator of 1. For example, 5 would be written as 5/1.

To graph a line given an equation in slope-intercept form, the y-intercept should first be plotted. For example, to graph $y = -\frac{2}{3}x + 7$, the y-intercept of 7 would be plotted on the y-axis (vertical axis) at the point (0, 7). Next, the slope would be used to determine a second point for the line. Note that all that is necessary to graph a line is two points on that line. The slope will indicate how to get from one point on the line to another. The slope expresses vertical change (y) compared to horizontal change (x) and therefore is sometimes referred to as $\frac{rise}{run}$. The numerator indicates the change in the y value (move up for positive integers and move down for negative integers), and the denominator indicates the change in the x value. For the previous example, using the slope of $-\frac{2}{3}$, from the first point at the y-intercept, the second point should be found by counting down 2 and to the right 3. This point would be located at (3, 5).

Graphing a Line in Standard Form

When an equation is written in standard form, $Ax + By = C$, it is easy to identify the x- and y-intercepts for the graph of the line. Just as the y-intercept is the point at which the line intercepts the y-axis, the x-intercept is the point at which the line intercepts the x-axis. At the y-intercept, $x = 0$; and at the x-intercept, $y = 0$. Given an equation in standard form, $x = 0$ should be used to find the y-intercept. Likewise, $y = 0$ should be used to find the x-intercept. For example, to graph $3x + 2y = 6$, 0 for y results in $3x + 2(0) = 6$. Solving for y yields $x = 2$; therefore, an ordered pair for the line is (2, 0). Substituting 0 for x results in $3(0) + 2y = 6$. Solving for y yields $y = 3$; therefore, an ordered pair for the line is (0, 3). The two ordered pairs (the x- and y-intercepts) can be plotted and a straight line through them can be constructed.

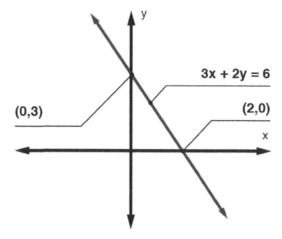

T - chart		Intercepts
x	**y**	**x** - intercept : (2,0)
0	3	**y** - intercept : (0,3)
2	0	

Writing the Equation of a Line Given its Graph

Given the graph of a line, its equation can be written in two ways. If the y-intercept is easily identified (is an integer), it and another point can be used to determine the slope. When determining $\frac{change\ in\ y}{change\ in\ x}$ from one point to another on the graph, the distance for $\frac{rise}{run}$ is being figured. The equation should be written in slope-intercept form, $y = mx + b$, with m representing the slope and b representing the y-intercept.

The equation of a line can also be written by identifying two points on the graph of the line. To do so, the slope is calculated and then the values are substituted for the slope and either of the ordered pairs into the point-slope form of an equation.

Vertical, Horizontal, Parallel, and Perpendicular Lines

For a vertical line, the value of x remains constant (for all ordered pairs (x, y) on the line, the value of x is the same); therefore, the equations for all vertical lines are written in the form $x = number$. For example, a vertical line that crosses the x-axis at -2 would have an equation of $x = -2$. For a horizontal line, the value of y remains constant; therefore, the equations for all horizontal lines are written in the form $y = number$.

Parallel lines extend in the same exact direction without ever meeting. Their equations have the same slopes and different y-intercepts. For example, given a line with an equation of $y = -3x + 2$, a parallel line would have a slope of -3 and a y-intercept of any value other than 2. Perpendicular lines intersect to form a right angle. Their equations have slopes that are opposite reciprocal (the sign is changed and the fraction is flipped; for example, $-\frac{2}{3}$ and $\frac{3}{2}$) and y-intercepts that may or may not be the same. For

example, given a line with an equation of $y = \frac{1}{2}x + 7$, a perpendicular line would have a slope of $-\frac{2}{1}$ and any value for its y-intercept.

Graphing Solution Sets for Systems of Linear Inequalities in Two Variables

The solution set for a system of inequalities is the region of a graph consisting of ordered pairs that make both inequalities true. To graph the solution set, each linear inequality should first be graphed with appropriate shading. The region of the graph should be identified where the shading for the two inequalities overlaps. This region contains the solution set for the system.

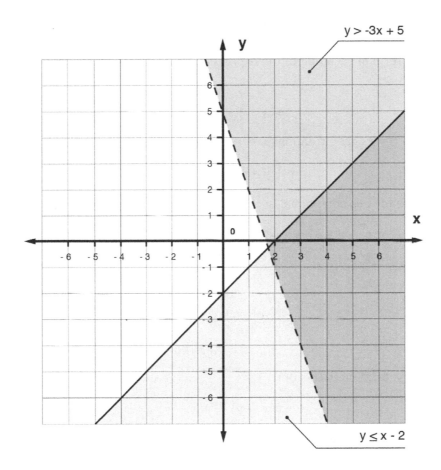

An ordered pair from the region of solutions can be selected to test in the system of inequalities.

Just as with manipulating linear inequalities in one variable, if dividing by a negative number in working with a linear inequality in two variables, the relationship reverses and the inequality sign should be flipped.

Factoring

Factors for polynomials are similar to factors for integers—they are numbers, variables, or polynomials that, when multiplied together, give a product equal to the polynomial in question. One polynomial is a factor of a second polynomial if the second polynomial can be obtained from the first by multiplying by a third polynomial.

$6x^6 + 13x^4 + 6x^2$ can be obtained by multiplying together $(3x^4 + 2x^2)(2x^2 + 3)$. This means $2x^2 + 3$ and $3x^4 + 2x^2$ are factors of $6x^6 + 13x^4 + 6x^2$.

In general, finding the factors of a polynomial can be tricky. However, there are a few types of polynomials that can be factored in a straightforward way.

If a certain monomial is in each term of a polynomial, it can be factored out. There are several common forms polynomials take, which if you recognize, you can solve. The first example is a perfect square trinomial. To factor this polynomial, first expand the middle term of the expression:

$$x^2 + 2xy + y^2$$

$$x^2 + xy + xy + y^2$$

Factor out a common term in each half of the expression (in this case x from the left and y from the right):

$$x(x + y) + y(x + y)$$

Then the same can be done again, treating $(x + y)$ as the common factor:

$$(x + y)(x + y) = (x + y)^2$$

Therefore, the formula for this polynomial is:

$$x^2 + 2xy + y^2 = (x + y)^2$$

Next is another example of a perfect square trinomial. The process is the similar, but notice the difference in sign:

$$x^2 - 2xy + y^2$$

$$x^2 - xy - xy + y^2$$

Factor out the common term on each side:

$$x(x - y) - y(x - y)$$

Factoring out the common term again:

$$(x - y)(x - y) = (x - y)^2$$

Thus:

$$x^2 - 2xy + y^2 = (x - y)^2$$

The next is known as a difference of squares. This process is effectively the reverse of binomial multiplication:

$$x^2 - y^2$$

$$x^2 - xy + xy - y^2$$

$$x(x - y) + y(x - y)$$

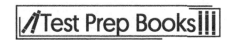

$$(x + y)(x - y)$$

Therefore:

$$x^2 - y^2 = (x + y)(x - y)$$

The following two polynomials are known as the sum or difference of cubes. These are special polynomials that take the form of $x^3 + y^3$ or $x^3 - y^3$. The following formula factors the sum of cubes:

$$x^3 + y^3 = (x + y)(x^2 - xy + y^2)$$

Next is the difference of cubes, but note the change in sign. The formulas for both are similar, but the order of signs for factoring the sum or difference of cubes can be remembered by using the acronym SOAP, which stands for "same, opposite, always positive." The first sign is the same as the sign in the first expression, the second is opposite, and the third is always positive. The next formula factors the difference of cubes:

$$x^3 - y^3 = (x - y)(x^2 + xy + y^2)$$

The following two examples are expansions of cubed binomials. Similarly, these polynomials always follow a pattern:

$$x^3 + 3x^2y + 3xy^2 + y^3 = (x + y)^3$$
$$x^3 - 3x^2y + 3xy^2 - y^3 = (x - y)^3$$

These rules can be used in many combinations with one another. For example, the expression $3x^3 - 24$ has a common factor of 3, which becomes:

$$3(x^3 - 8)$$

A difference of cubes still remains which can then be factored out:

$$3(x - 2)(x^2 + 2x + 4)$$

There are no other terms to be pulled out, so this expression is completely factored.

When factoring polynomials, a good strategy is to multiply the factors to check the result. Let's try another example:

$$4x^3 + 16x^2$$

Both sides of the expression can be divided by 4, and both contain x^2, because $4x^3$ can be thought of as $4x^2(x)$, so the common term can simply be factored out:

$$4x^2(x + 4)$$

It sometimes can be necessary to rewrite the polynomial in some clever way before applying the above rules. Consider the problem of factoring $x^4 - 1$. This does not immediately look like any of the previous polynomials. However, it's possible to think of this polynomial as $x^4 - 1 = (x^2)^2 - (1^2)^2$, and now it can be treated as a difference of squares to simplify this:

$$(x^2)^2 - (1^2)^2$$

$$(x^2)^2 - x^2 1^2 + x^2 1^2 - (1^2)^2$$

$$x^2(x^2 - 1^2) + 1^2(x^2 - 1^2)$$

$$(x^2 + 1^2)(x^2 - 1^2)$$

$$(x^2 + 1)(x^2 - 1)$$

Factoring an Algebraic Expression

There are many different ways to write algebraic expressions and equations that are equivalent to each other. Converting expressions from standard form to factored form and vice versa are skills commonly used in advanced mathematics. Standard form of an expression arranges terms with variables powers in descending order (highest exponent to lowest and then constants). Factored form displays an expression as the product of its factors (what can be multiplied to produce the expression).

Converting Standard Form to Factored Form

To factor an expression, a greatest common factor needs to be factored out first. Then, if possible, the remaining expression needs to be factored into the product of binomials. A binomial is an expression with two terms.

Greatest Common Factor

The greatest common factor (GCF) of a monomial (one term) consists of the largest number that divides evenly into all coefficients (number part of a term), and if all terms contain the same variable, the variable with the lowest exponent. The GCF of $3x^4 - 9x^3 + 12x^2$ would be $3x^2$. To write the factored expression, every term needs to be divided by the GCF, then the product of the resulting quotient and the GCF (using parentheses to show multiplication) should be written. For the previous example, the factored expression would be $3x^2(x^2 - 3x + 4)$.

Factoring $Ax^2 + Bx + C$ When A = 1

To factor a quadratic expression in standard form when the value of a is equal to 1, the factors that multiply to equal the value of c should be found and then added to equal the value of b (the signs of b and c should be included). The factored form for the expression will be the product of binomials: $(x + factor1)(x + factor2)$. Here's a sample expression: $x^2 - 4x - 5$. The two factors that multiply to equal c(-5) and add together to equal b(-4) are -5 and 1. Therefore, the factored expression would be $(x - 5)(x + 1)$. Note $(x + 1)(x - 5)$ is equivalent.

Factoring a Difference of Squares

A difference of squares is a binomial expression where both terms are perfect squares (perfect square-perfect square). Perfect squares include 1, 4, 9, 16 . . . and x^2, x^4, x^6 . . .

The factored form of a difference of squares will be:

$$(\sqrt{term1} + \sqrt{term2})(\sqrt{term1} - \sqrt{term2})$$

For example:

$$x^2 - 4 = (x + 2)(x - 2)$$

And

$$25x^6 - 81 = (5x^3 + 9)(5x^3 - 9)$$

Factoring Ax² + Bx + C when A ≠ 1

To factor a quadratic expression in standard form when the value of a is not equal to 1, the factors that multiply to equal the value of $a \times c$ should be found and then added to equal the value of b. Next, the expression splitting the bx term should be rewritten using those factors. Instead of three terms, there will now be four. Then the first two terms should be factored using GCF, and a common binomial should be factored from the last two terms. The factored form will be: (common binomial) (2 terms out of binomials). In the sample expression $2 \times 2 + 11x + 12$, the value of $a \times c$ ($2x12$) $= 24$. Two factors that multiply to 24 and added together to yield b(11) are 8 and 3. The bx term (11x) can be rewritten by splitting it into the factors: $2 \times 2 + 8x + 3x + 12$. A GCF from the first two terms can be factored as: $2x(x + 4) + 3x + 12$. A common binomial from the last two terms can then be factored as: $2(x + 4) + 3(x + 4)$. The factored form can be written as a product of binomials: $(x + 4)(2x + 3)$.

Converting Factored Form to Standard Form

To convert an expression from factored form to standard form, the factors are multiplied.

Quadratics

Quadratic Equations

A **quadratic equation** can be written in the form $y = ax^2 + bx + c$. The u-shaped graph of a quadratic function is called a **parabola**. The graph can either open up or open down (upside down u). The graph is symmetric about a vertical line, called the axis of symmetry. Corresponding points on the parabola are directly across from each other (same y-value) and are the same distance from the axis of symmetry (on either side). The axis of symmetry intersects the parabola at its vertex. The y-value of the vertex represents the minimum or maximum value of the function. If the graph opens up, the value of a in its

equation is positive, and the vertex represents the minimum of the function. If the graph opens down, the value of a in its equation is negative, and the vertex represents the maximum of the function.

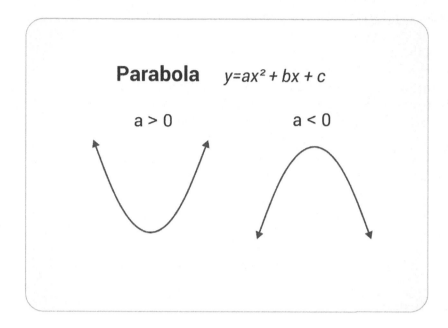

For a quadratic equation where the value of a is positive, as the inputs increase, the outputs increase until a certain value (maximum of the function) is reached. As inputs increase past the value that corresponds with the maximum output, the relationship reverses, and the outputs decrease. For a quadratic equation where a is negative, as the inputs increase, the outputs (1) decrease, (2) reach a maximum, and (3) then increase.

Consider a ball thrown straight up into the air. As time passes, the height of the ball increases until it reaches its maximum height. After reaching the maximum height, as time increases, the height of the ball decreases (it is falling toward the ground). This relationship can be expressed as a quadratic function where time is the input (x) and the height of the ball is the output (y).

To write a quadratic equation that models a real-life scenario, the following is needed: its vertex and any other ordered pair; or any three ordered pairs for the function. Given three ordered pairs, they should be substituted into the general form ($y = ax^2 + bx + c$) to create a system of three equations. For example, given the ordered pairs (2, 3), (3, 13), and (4, 29), it yields:

$$3 = a(2)2 + b(2) + c \rightarrow 4a + 2b + c = 3$$

$$13 = a(3)2 + b(3) + c \rightarrow 9a + 3b + c = 13$$

$$29 = a(4)2 + b(4) + c \rightarrow 16a + 24b + c = 29$$

The values for a, b, and c in the system can be found and substituted into the general form to write the equation of the function. In this case, the equation is $y = 3x^2 - 5x + 1$.

Given a curve of best-fit that models a quadratic relationship, the equation of the parabola can be written by identifying the vertex of the parabola and another point on the graph. The values for the vertex (h, k) and the point (x, y) should be substituted into the vertex form of a quadratic function, $y =$

$a(x - h)^2 + k$, to determine the value of a. To write the equation of a quadratic model with a vertex of (4, 7) and containing the point (8, 3), the values for h, k, x, and y should be substituted into the vertex form of a quadratic equation, resulting in $3 = a(8 - 4)^2 + 7$. Solving for a, yields $a = -\frac{1}{4}$. Therefore, the equation can be written as $y = -\frac{1}{4}(x - 4)^2 + 7$. The vertex form can be manipulated in order to write the quadratic equation in standard form.

Solving a Quadratic Equation

A quadratic equation is one in which the highest exponent of the variable is 2. A quadratic equation can have two, one, or zero real solutions. Depending on its structure, a quadratic equation can be solved by (1) factoring, (2) taking square roots, or (3) using the quadratic formula.

Solving Quadratic Equations by Factoring

To solve a quadratic equation by factoring, the equation should first be manipulated to set the quadratic expression equal to zero. Next, the quadratic expression should be factored using the appropriate method(s). Then each factor should be set equal to zero. If two factors multiply to equal zero, then one or both factors must equal zero. Finally, each equation should be solved. Here's a sample: $x^2 - 10 = 3x - 6$. The expression should be set equal to zero: $x^2 - 3x - 4 = 0$. The expression should be factored: $(x - 4)(x + 1) = 0$. Each factor should be set equal to zero: $x - 4 = 0$; $x + 1 = 0$. Solving yields $x = 4$ or $x = -1$.

Solving Quadratic Equations by Taking Square Roots

If a quadratic equation does not have a linear term (variable to the first power), it can be solved by taking square roots. This means x^2 needs to be isolated and then the square root of both sides of the equation should be isolated. There will be two solutions because square roots can be positive or negative. ($\sqrt{4}$ = 2 or -2 because $2 \times 2 = 4$ and $-2 \times -2 = 4$.) Here's a sample equation: $3x^2 - 12 = 0$. Isolating x^2 yields $x^2 = 4$. The square root of both sides is then solved: $x = 2$ or -2.

The Quadratic Formula

When a quadratic expression cannot be factored or is difficult to factor, the quadratic formula can be used to solve the equation. To do so, the equation must be in the form $ax^2 + bx + c = 0$. The quadratic formula is:

$$x = \frac{-b \pm \sqrt{b^2 - 4ac}}{2a}$$

(The $\pm$ symbol indicates that two calculations are necessary, one using + and one using −.) Here's a sample equation: $3x^2 - 2x = 3x + 2$. First, the quadratic expression should be set equal to zero: $3x^2 - 5x - 2 = 0$. Then the values are substituted for a(3), b(-5), and c(-2) into the formula:

$$x = \frac{-(-5) \pm \sqrt{(-5)^2 - 4(3)(-2)}}{2(3)}$$

Simplification yields:

$$x = \frac{5 \pm \sqrt{49}}{6} \rightarrow x = \frac{5 \pm 7}{6}$$

Calculating two values for x using $+$ and $-$ yields:

$$x = \frac{5 + 7}{6}; x = \frac{5 - 7}{6}$$

Simplification yields:

$$x = 2 \text{ or } -\frac{1}{3}.$$

Just as with any equation, solutions should be checked by substituting the value into the original equation.

Solving a System of One Linear Equation and One Quadratic Equation

A system of equations consists of two variables in two equations. A solution to the system is an ordered pair (x, y) that makes both equations true. When displayed graphically, a solution to a system is a point of intersection between the graphs of the equations. When a system consists of one linear equation and one quadratic equation, there may be one, two, or no solutions. If the line and parabola intersect at two points, there are two solutions to the system; if they intersect at one point, there is one solution; if they do not intersect, there is no solution.

Systems with One Linear Equation and One Quadratic Equation

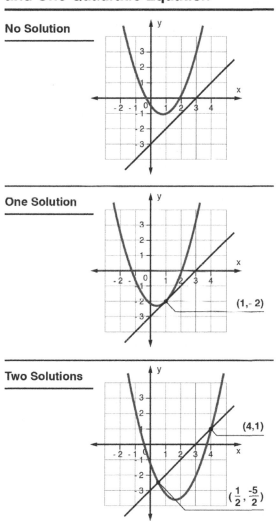

One method for solving a system of one linear equation and one quadratic equation is to graph both functions and identify point(s) of intersection. This, however, is not always practical. Graph paper may not be available, or the intersection points may not be easily identified. Solving the system algebraically involves using the substitution method. Consider the following system: $y = x^2 + 9x + 11; y = 2x - 1$. The equivalent value of y should be substituted from the linear equation ($2x - 1$) into the quadratic equation. The resulting equation is $2x - 1 = x^2 + 9x + 11$. Next, this quadratic equation should be

119

solved using the appropriate method: factoring, taking square roots, or using the quadratic formula. Solving this quadratic equation by factoring results in $x = -4$ or $x = -3$. Next, the corresponding y-values should be found by substituting the x-values into the original linear equation: $y = 2(-4) - 1; y = 2(-3) - 1$. The solutions should be written as ordered pairs: (-4, -9) and (-3, -7). Finally, the possible solutions should be checked by substituting each into both of the original equations. In this case, both solutions "check out."

Functions

A **function** is defined as a relationship between inputs and outputs where there is only one output value for a given input. As an example, the following function is in function notation: $f(x) = 3x - 4$. The $f(x)$ represents the output value for an input of x. If $x = 2$, the equation becomes:

$$f(2) = 3(2) - 4 = 6 - 4 = 2$$

The input of 2 yields an output of 2, forming the ordered pair $(2, 2)$. The following set of ordered pairs corresponds to the given function: $(2, 2), (0, -4), (-2, -10)$. The set of all possible inputs of a function is its **domain**, and all possible outputs is called the **range**. By definition, each member of the domain is paired with only one member of the range.

Functions can also be defined recursively. In this form, they are not defined explicitly in terms of variables. Instead, they are defined using previously-evaluated function outputs, starting with either $f(0)$ or $f(1)$. An example of a recursively-defined function is:

$$f(1) = 2, f(n) = 2f(n - 1) + 2n, n > 1$$

The domain of this function is the set of all integers.

Creating Functions Using Function Notation

A **function** is defined as a relationship between inputs and outputs where there is only one output value for a given input. As an example, the following function is in function notation:

$$f(x) = 3x - 4$$

The $f(x)$ represents the output value for an input of x. If $x = 2$, the equation becomes:

$$f(2) = 3(2) - 4 = 6 - 4 = 2$$

The input of 2 yields an output of 2, forming the ordered pair $(2, 2)$. The following set of ordered pairs corresponds to the given function:

$$(2, 2), (0, -4), (-2, -10)$$

The set of all possible inputs of a function is its **domain**, and all possible outputs is called the **range**. By definition, each member of the domain is paired with only one member of the range.

Functions can also be defined recursively. In this form, they are not defined explicitly in terms of variables. Instead, they are defined using previously-evaluated function outputs, starting with either $f(0)$ or $f(1)$. An example of a recursively-defined function is:

$$f(1) = 2, f(n) = 2f(n - 1) + 2n, n > 1$$

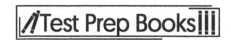

The domain of this function is the set of all integers.

A function $f(x)$ is a mathematical object which takes one number, x, as an input and gives a number in return. The input is called the **independent variable**. If the variable is set equal to the output, as in $y = f(x)$, then this is called the **dependent variable**. To indicate the dependent value a function, y, gives for a specific independent variable, x, the notation y = $f(x)$ is used.

The **domain** of a function is the set of values that the independent variable is allowed to take. Unless otherwise specified, the domain is any value for which the function is well defined. The **range** of the function is the set of possible outputs for the function.

In many cases, a function can be defined by giving an equation. For instance, $f(x) = x^2$ indicates that given a value for x, the output of f is found by squaring x.

Not all equations in x and y can be written in the form $y = f(x)$. An equation can be written in such a form if it satisfies the **vertical line test**: no vertical line meets the graph of the equation at more than a single point. In this case, y is said to be a *function of x*. If a vertical line meets the graph in two places, then this equation cannot be written in the form $y = f(x)$.

The graph of a function $f(x)$ is the graph of the equation $y = f(x)$. Thus, it is the set of all pairs (x, y) where $y = f(x)$. In other words, it is all pairs $(x, f(x))$. The x-intercepts are called the **zeros** of the function. The y-intercept is given by $f(0)$.

If, for a given function f, the only way to get $f(a) = f(b)$ is for $a = b$, then f is *one-to-one*. Often, even if a function is not one-to-one on its entire domain, it is one-to-one by considering a restricted portion of the domain.

A function $f(x) = k$ for some number k is called a **constant function**. The graph of a constant function is a horizontal line.

The function $f(x) = x$ is called the **identity function**. The graph of the identity function is the diagonal line pointing to the upper right at 45 degrees, $y = x$.

A function is called **monotone** if it is either always increasing or always decreasing. For example, the functions $f(x) = 3x$ and $f(x) = -x^5$ are monotone.

An **even function** looks the same when flipped over the *y*-axis: $f(x) = f(-x)$. The following image shows a graphic representation of an even function.

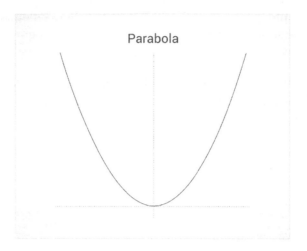

Parabola

An **odd function** looks the same when flipped over the *y*-axis and then flipped over the *x*-axis: $f(x) = -f(-x)$. The following image shows an example of an odd function.

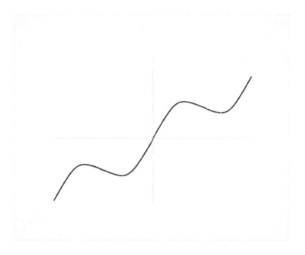

Domain and Range

The domain and range of a function can be found visually by its plot on the coordinate plane. In the function $f(x) = x^2 - 3$, for example, the domain is all real numbers because the parabola stretches as far left and as far right as it can go, with no restrictions. This means that any input value from the real number system will yield an answer in the real number system. For the range, the inequality $y \geq -3$ would be used to describe the possible output values because the parabola has a minimum at $y = -3$. This means there will not be any real output values less than -3 because -3 is the lowest value it reaches on the y-axis.

These same answers for domain and range can be found by observing a table. The table below shows that from input values $x = -1$ to $x = 1$, the output results in a minimum of -3. On each side of $x = 0$, the numbers increase, showing that the range is all real numbers greater than or equal to -3.

x (domain/input)	y (range/output)
-2	1
-1	-2
0	-3
-1	-2
2	1

Function Behavior

Different types of functions behave in different ways. A function is defined to be increasing over a subset of its domain if for all $x_1 \geq x_2$ in that interval, $f(x_1) \geq f(x_2)$. Also, a function is decreasing over an interval if for all $x_1 \geq x_2$ in that interval, $f(x_1) \leq f(x_2)$. A point in which a function changes from increasing to decreasing can also be labeled as the **maximum value** of a function if it is the largest point the graph reaches on the y-axis. A point in which a function changes from decreasing to increasing can be labeled as the minimum value of a function if it is the smallest point the graph reaches on the y-axis. Maximum values are also known as **extreme values**. The graph of a continuous function does not have any breaks or jumps in the graph. This description is not true of all functions. A radical function, for example, $f(x) = \sqrt{x}$, has a restriction for the domain and range because there are no real negative inputs or outputs for this function. The domain can be stated as $x \geq 0$, and the range is $y \geq 0$.

A piecewise-defined function also has a different appearance on the graph. In the following function, there are three equations defined over different intervals. It is a function because there is only one y-value for each x-value, passing the Vertical Line Test. The domain is all real numbers less than or equal to 6. The range is all real numbers greater than zero. From left to right, the graph decreases to zero, then increases to almost 4, and then jumps to 6.

From input values greater than 2, the input decreases just below 8 to 4, and then stops.

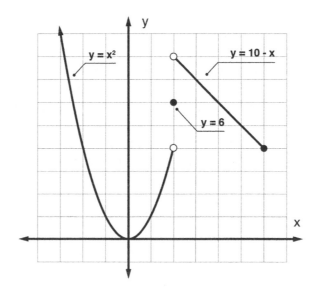

Logarithmic and exponential functions also have different behavior than other functions. These two types of functions are inverses of each other. The **inverse** of a function can be found by switching the place of x and y, and solving for y. When this is done for the exponential equation, $y = 2^x$, the function $y = \log_2 x$ is found. The general form of a **logarithmic function** is $y = \log_b x$, which says b raised to the y power equals x.

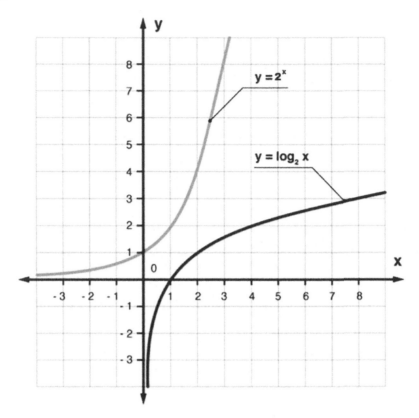

The thick black line on the graph above represents the logarithmic function:

$$y = \log_2 x$$

This curve passes through the point $(1, 0)$, just as all log functions do, because any value $b^0 = 1$. The graph of this logarithmic function starts very close to zero but does not touch the y-axis. The output value will never be zero by the definition of logarithms. The thinner gray line seen above represents the exponential function $y = 2^x$. The behavior of this function is opposite the logarithmic function because the graph of an inverse function is the graph of the original function flipped over the line $y = x$. The curve passes through the point $(0, 1)$ because any number raised to the zero power is one. This curve also gets very close to the x-axis but never touches it because an exponential expression never has an output of zero. The x-axis on this graph is called a horizontal asymptote. An **asymptote** is a line that represents a boundary for a function. It shows a value that the function will get close to, but never reach.

Functions can also be described as being even, odd, or neither. If $f(-x) = f(x)$, the function is even. For example, the function $f(x) = x^2 - 2$ is even. Plugging in $x = 2$ yields an output of $y = 2$. After changing the input to $x = -2$, the output is still $y = 2$. The output is the same for opposite inputs. Another way to observe an even function is by the symmetry of the graph. If the graph is symmetrical

about the axis, then the function is even. If the graph is symmetric about the origin, then the function is odd. Algebraically, if $f(-x) = -f(x)$, the function is odd.

Also, a function can be described as **periodic** if it repeats itself in regular intervals. Common periodic functions are trigonometric functions. For example, $y = \sin x$ is a periodic function with period 2π because it repeats itself every 2π units along the x-axis.

Common Functions

Three common functions used to model different relationships between quantities are linear, quadratic, and exponential functions. **Linear functions** are the simplest of the three, and the independent variable x has an exponent of 1. Written in the most common form, $y = mx + b$, the coefficient of x tells how fast the function grows at a constant rate, and the b-value tells the starting point. A **quadratic** function has an exponent of 2 on the independent variable x. Standard form for this type of function is $y = ax^2 + bx + c$, and the graph is a parabola. These type functions grow at a changing rate. An **exponential** function has an independent variable in the exponent $y = ab^x$. The graph of these types of functions is described as **growth** or **decay**, based on whether the base, b, is greater than or less than 1. These functions are different from quadratic functions because the base stays constant. A common base is base e.

The following three functions model a linear, quadratic, and exponential function respectively: $y = 2x$, $y = x^2$, and $y = 2^x$. Their graphs are shown below. The first graph, modeling the linear function, shows that the growth is constant over each interval. With a horizontal change of 1, the vertical change is 2. It models a constant positive growth. The second graph shows the quadratic function, which is a curve that is symmetric across the y-axis. The growth is not constant, but the change is mirrored over the axis. The last graph models the exponential function, where the horizontal change of 1 yields a vertical change that increases more and more. The exponential graph gets very close to the x-axis, but never touches it, meaning there is an asymptote there. The y-value can never be zero because the base of 2 can never be raised to an input value that yields an output of zero.

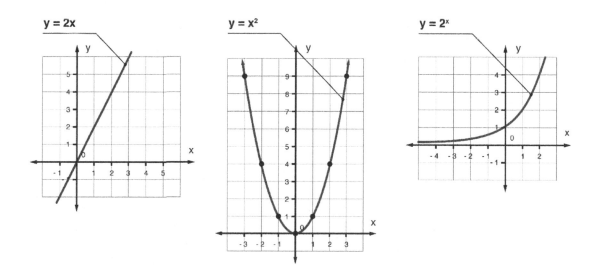

The three tables below show specific values for three types of functions. The third column in each table shows the change in the y-values for each interval. The first table shows a constant change of 2 for each equal interval, which matches the slope in the equation $y = 2x$. The second table shows an increasing change, but it also has a pattern. The increase is changing by 2 more each time, so the change is quadratic. The third table shows the change as factors of the base, 2. It shows a continuing pattern of factors of the base.

y = 2x				y = x²				y = 2ˣ		
x	y	Δy		x	y	Δy		x	y	Δy
1	2			1	1			1	2	
2	4	2		2	4	3		2	4	2
3	6	2		3	9	5		3	8	4
4	8	2		4	16	7		4	16	8
5	10	2		5	25	9		5	32	16

Given a table of values, the type of function can be determined by observing the change in y over equal intervals. For example, the tables below model two functions. The changes in interval for the x-values is 1 for both tables. For the first table, the y-values increase by 5 for each interval. Since the change is constant, the situation can be described as a linear function. The equation would be $y = 5x + 3$. For the second table, the change for y is 5, 20, 100, and 500, respectively. The increases are multiples of 5, meaning the situation can be modeled by an exponential function. The equation $y = 5^x + 3$ models this situation.

x	y
0	3
1	8
2	13
3	18
4	23

x	y
0	3
1	8
2	28
3	128
4	628

Quadratic equations can be used to model real-world area problems. For example, a farmer may have a rectangular field that he needs to sow with seed. The field has length $x + 8$ and width $2x$. The formula for area should be used: $A = lw$. Therefore:

$$A = (x + 8) \times 2x = 2x^2 + 16x$$

The possible values for the length and width can be shown in a table, with input x and output A. If the equation was graphed, the possible area values can be seen on the y-axis for given x-values.

Exponential growth and decay can be found in real-world situations. For example, if a piece of notebook paper is folded 25 times, the thickness of the paper can be found. To model this situation, a table can be used. The initial point is one-fold, which yields a thickness of 2 papers. For the second fold, the thickness is 4. Since the thickness doubles each time, the table below shows the thickness for the next few folds. Notice the thickness changes by the same factor each time. Since this change for a constant interval of

folds is a factor of 2, the function is exponential. The equation for this is $y = 2^x$. For twenty-five folds, the thickness would be 33,554,432 papers.

x (folds)	y (paper thickness)
0	1
1	2
2	4
3	8
4	16
5	32

One exponential formula that is commonly used is the **interest formula**: $A = Pe^{rt}$. In this formula, interest is compounded continuously. A is the value of the investment after the time, t, in years. P is the initial amount of the investment, r is the interest rate, and e is the constant equal to approximately 2.718. Given an initial amount of $200 and a time of 3 years, if interest is compounded continuously at a rate of 6%, the total investment value can be found by plugging each value into the formula. The invested value at the end is $239.44. In more complex problems, the final investment may be given, and the rate may be the unknown. In this case, the formula becomes $239.44 = 200e^{r3}$. Solving for r requires isolating the exponential expression on one side by dividing by 200, yielding the equation $1.20 = e^{r3}$. Taking the natural log of both sides results in $\ln(1.2) = r3$. Using a calculator to evaluate the logarithmic expression, $r = 0.06 = 6\%$.

When working with logarithms and exponential expressions, it is important to remember the relationship between the two. In general, the logarithmic form is $y = log_b x$ for an exponential form $b^y = x$. Logarithms and exponential functions are inverses of each other.

Interpreting and Building a Function Within a Context

Functions can be built out of the context of a situation. For example, the relationship between the money paid for a gym membership and the months that someone has been a member can be described through a function. If the one-time membership fee is $40 and the monthly fee is $30, then the function can be written $f(x) = 30x + 40$. The x-value represents the number of months the person has been part of the gym, while the output is the total money paid for the membership. The table below shows this relationship. It is a representation of the function because the initial cost is $40 and the cost increases each month by $30.

x (months)	y (money paid to gym)
0	40
1	70
2	100
3	130

Functions can also be built from existing functions. For example, a given function $f(x)$ can be transformed by adding a constant, multiplying by a constant, or changing the input value by a constant. The new function $g(x) = f(x) + k$ represents a vertical shift of the original function. In $f(x) = 3x - 2$, a vertical shift 4 units up would be:

$$g(x) = 3x - 2 + 4 = 3x + 2$$

Multiplying the function times a constant k represents a vertical stretch, based on whether the constant is greater than or less than 1. The function

$$g(x) = kf(x) = 4(3x - 2) = 12x - 8$$

represents a stretch. Changing the input x by a constant forms the function:

$$g(x) = f(x + k) = 3(x + 4) - 2 = 3x + 12 - 2 = 3x + 10$$

and this represents a horizontal shift to the left 4 units. If $(x - 4)$ was plugged into the function, it would represent a vertical shift.

A composition function can also be formed by plugging one function into another. In function notation, this is written:

$$(f \circ g)(x) = f(g(x))$$

For two functions $f(x) = x^2$ and $g(x) = x - 3$, the composition function becomes:

$$f(g(x)) = (x - 3)^2 = x^2 - 6x + 9$$

The composition of functions can also be used to verify if two functions are inverses of each other. Given the two functions $f(x) = 2x + 5$ and $g(x) = \frac{x-5}{2}$, the composition function can be found $(f \circ g)(x)$. Solving this equation yields:

$$f(g(x)) = 2\left(\frac{x - 5}{2}\right) + 5 = x - 5 + 5 = x$$

It also is true that $g(f(x)) = x$. Since the composition of these two functions gives a simplified answer of x, this verifies that $f(x)$ and $g(x)$ are inverse functions. The domain of $f(g(x))$ is the set of all x-values in the domain of $g(x)$ such that $g(x)$ is in the domain of $f(x)$. Basically, both $f(g(x))$ and $g(x)$ have to be defined.

To build an inverse of a function, $f(x)$ needs to be replaced with y, and the x and y values need to be switched. Then, the equation can be solved for y. For example, given the equation $y = e^{2x}$, the inverse can be found by rewriting the equation $x = e^{2y}$. The natural logarithm of both sides is taken down, and the exponent is brought down to form the equation:

$$\ln(x) = \ln(e)\, 2y$$

ln(e)=1, which yields the equation $\ln(x) = 2y$. Dividing both sides by 2 yields the inverse equation

$$\frac{\ln(x)}{2} = y = f^{-1}(x)$$

The domain of an inverse function is the range of the original function, and the range of an inverse function is the domain of the original function. Therefore, an ordered pair (x, y) on either a graph or a table corresponding to $f(x)$ means that the ordered pair (y, x) exists on the graph of $f^{-1}(x)$. Basically, if $f(x) = y$, then $f^{-1}(y) = x$. For a function to have an inverse, it must be one-to-one. That means it must pass the **Horizontal Line Test**, and if any horizontal line passes through the graph of the function twice, a function is not one-to-one. The domain of a function that is not one-to-one can be restricted to an interval in which the function is one-to-one, to be able to define an inverse function.

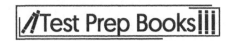

Functions can also be formed from combinations of existing functions.

Given $f(x)$ and $g(x)$, the following can be built:

$$f + g$$

$$f - g$$

$$fg$$

$$\frac{f}{g}$$

The domains of $f + g, f - g$, and fg are the intersection of the domains of f and g. The domain of $\frac{f}{g}$ is the same set, excluding those values that make $g(x) = 0$.

For example, if:

$$f(x) = 2x + 3$$

$$g(x) = x + 1$$

then

$$\frac{f}{g} = \frac{2x + 3}{x + 1}$$

Its domain is all real numbers except -1.

Radical and Rational Equations

Radical Equations

Equations with radicals containing numbers only as the radicand are solved the same way that an equation without a radical would be. For example, $3x + \sqrt{81} = 45$ would be solved using the same steps as if solving $2x + 4 = 12$. Radical equations are those in which the variable is part of the radicand. For example, $\sqrt{5x + 1} - 6 = 0$ and $\sqrt{x - 3} + 5 = x$ would be considered radical equations.

To solve a radical equation, the radical should be isolated and both sides of the equation should be raised to the same power to cancel the radical. Raising both sides to the second power will cancel a square root, raising to the third power will cancel a cube root, etc. To solve $\sqrt{5x + 1} - 6 = 0$, the radical should be isolated first: $\sqrt{5x + 1} = 6$. Then both sides should be raised to the second power: $(\sqrt{5x + 1})^2 = (6)^2 \rightarrow 5x + 1 = 36$. Lastly, the linear equation should be solved: $x = 7$.

If a radical equation contains a variable in the radicand and a variable outside of the radicand, it must be checked for extraneous solutions. An extraneous solution is one obtained by following the proper process for solving an equation but does not "check out" when substituted into the original equation. Here's a sample equation: $\sqrt{x - 3} + 5 = x$. Isolating the radical yields $\sqrt{x - 3} = x - 5$. Next, both sides should be squared to cancel the radical: $(\sqrt{x - 3})^2 = (x - 5)^2 \rightarrow x - 3 = (x - 5)(x - 5)$. The binomials should be multiplied: $x - 3 = x^2 - 10x + 25$. The quadratic equation is then solved:

$$0 = x^2 - 11x + 28 \rightarrow 0 = (x-7)(x-4) \rightarrow x - 7 = 0; x - 4 = 0 \rightarrow x = 7 \text{ or } x = 4$$

To check for extraneous solutions, each answer can be substituted, one at a time, into the original equation. Substituting 7 for x, results in $7 = 7$. Therefore, 7 is a solution. Substituting 4 for x results in $6 = 4$. This is false; therefore, 4 is an extraneous solution.

Rational Expressions, Equations, and Functions

Rational Expressions

A **rational expression** is a fraction where the numerator and denominator are both polynomials. Some examples of rational expressions include the following: $\frac{4x^3y^5}{3z^4}$, $\frac{4x^3+3x}{x^2}$, and $\frac{x^2+7x+10}{x+2}$. Since these refer to expressions and not equations, they can be simplified but not solved. Exponents on the variables are restricted to whole numbers, which means roots and negative exponents are not included in rational expressions.

Rational expressions can be transformed by factoring. For example, the expression $\frac{x^2-5x+6}{(x-3)}$ can be rewritten by factoring the numerator to obtain:

$$\frac{(x-3)(x-2)}{(x-3)}$$

Therefore, the common binomial $(x-3)$ can cancel so that the simplified expression is:

$$\frac{(x-2)}{1} = (x-2)$$

Additionally, other rational expressions can be rewritten to take on different forms. Some may be factorable in themselves, while others can be transformed through arithmetic operations. Rational expressions are closed under addition, subtraction, multiplication, and division by a nonzero expression. **Closed** means that if any one of these operations is performed on a rational expression, the result will still be a rational expression. The set of all real numbers is another example of a set closed under all four operations.

Adding and subtracting rational expressions is based on the same concepts as adding and subtracting simple fractions. For both concepts, the denominators must be the same for the operation to take place. For example, here are two rational expressions:

$$\frac{x^3 - 4}{(x-3)} + \frac{x+8}{(x-3)}$$

Since the denominators are both $(x-3)$, the numerators can be combined by collecting like terms to form:

$$\frac{x^3 + x + 4}{(x-3)}$$

If the denominators are different, they need to be made common (the same) by using the Least Common Denominator (LCD). Each denominator needs to be factored, and the LCD contains each factor that appears in any one denominator the greatest number of times it appears in any denominator. The original expressions need to be multiplied times a form of 1, which will turn each denominator into the LCD. This process is like adding fractions with unlike denominators. It is also important when working with rational expressions to define what value of the variable makes the denominator zero. For this particular value, the expression is undefined.

Multiplication of rational expressions is performed like multiplication of fractions. The numerators are multiplied; then, the denominators are multiplied. The final fraction is then simplified. The expressions are simplified by factoring and cancelling out common terms. In the following example, the numerator of the second expression can be factored first to simplify the expression before multiplying:

$$\frac{x^2}{(x-4)} \times \frac{x^2 - x - 12}{2}$$

$$\frac{x^2}{(x-4)} \times \frac{(x-4)(x+3)}{2}$$

The $(x-4)$ on the top and bottom cancel out:

$$\frac{x^2}{1} \times \frac{(x+3)}{2}$$

Then multiplication is performed, resulting in:

$$\frac{x^3 + 3x^2}{2}$$

Dividing rational expressions is similar to the division of fractions, where division turns into multiplying by a reciprocal. So the following expression can be rewritten as a multiplication problem:

$$\frac{x^2 - 3x + 7}{x - 4} \div \frac{x^2 - 5x + 3}{x - 4}$$

$$\frac{x^2 - 3x + 7}{x - 4} \times \frac{x - 4}{x^2 - 5x + 3}$$

The $x - 4$ cancels out, leaving:

$$\frac{x^2 - 3x + 7}{x^2 - 5x + 3}$$

The final answers should always be completely simplified. If a function is composed of a rational expression, the zeros of the graph can be found from setting the polynomial in the numerator as equal to zero and solving. The values that make the denominator equal to zero will either exist on the graph as a hole or a vertical asymptote.

The following problem is an example of using rational expressions:

Reggie wants to lay sod in his rectangular backyard. The length of the yard is given by the expression $4x + 2$ and the width is unknown. The area of the yard is $20x + 10$. Reggie needs to find the width of the yard. Knowing that the area of a rectangle is length multiplied by width, an expression can be written to find the width: $\frac{20x+10}{4x+2}$, area divided by length. Simplifying this expression by factoring out 10 on the top and 2 on the bottom leads to this expression: $\frac{10(2x+1)}{2(2x+1)}$. Cancelling out the $2x + 1$ results in $\frac{10}{2} = 5$. The width of the yard is found to be 5 by simplifying the rational expression.

Rational Equations

A **rational equation** can be as simple as an equation with a ratio of polynomials, $\frac{p(x)}{q(x)}$, set equal to a value, where $p(x)$ and $q(x)$ are both polynomials. A rational equation has an equal sign, which is different from expressions. This leads to solutions, or numbers that make the equation true.

It is possible to solve rational equations by trying to get all of the x terms out of the denominator and then isolating them on one side of the equation. For example, to solve the equation $\frac{3x+2}{2x+3} = 4$, both sides get multiplied by $(2x + 3)$. This will cancel on the left side to yield: $3x + 2 = 4(2x + 3)$, then $3x + 2 = 8x + 12$. Now, subtract $8x$ from both sides, which yields $-5x + 2 = 12$. Subtracting 2 from both sides results in $-5x = 10$. Finally, both sides get divided by -5 to obtain $x = -2$.

Sometimes, when solving rational equations, it can be easier to try to simplify the rational expression by factoring the numerator and denominator first, then cancelling out common factors. For example, to solve $\frac{2x^2-8x+6}{x^2-3x+2} = 1$, the first step is to factor:

$$2x^2 - 8x + 6 = 2(x^2 - 4x + 3) = 2(x - 1)(x - 3)$$

Then, factor $x^2 - 3x + 2$ into $(x - 1)(x - 2)$. This turns the original equation into:

$$\frac{2(x - 1)(x - 3)}{(x - 1)(x - 2)} = 1$$

The common factor of $(x - 1)$ can be canceled, leaving:

$$\frac{2(x - 3)}{x - 1} = 1$$

Now the same method used in the previous example can be followed. Multiplying both sides by $x - 1$ and performing the multiplication on the left yields $2x - 6 = x - 1$, which can be simplified to $x = 5$.

Rational Functions

A **rational function** is similar to an equation, but it includes two variables. In general, a rational function is in the form: $f(x) = \frac{p(x)}{q(x)}$, where $p(x)$ and $q(x)$ are polynomials. Rational functions are defined everywhere except where the denominator is equal to zero. When the denominator is equal to zero, this indicates either a hole in the graph or an asymptote. An example of a function with an asymptote is shown below.

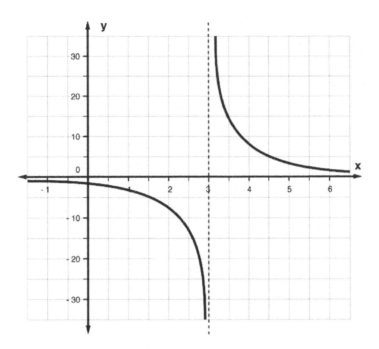

Polynomial Equations

Polynomials

An expression of the form ax^n, where n is a non-negative integer, is called a **monomial** because it contains one term. A sum of monomials is called a **polynomial**. For example, $-4x^3 + x$ is a polynomial, while $5x^7$ is a monomial. A function equal to a polynomial is called a **polynomial function**.

The monomials in a polynomial are also called the **terms** of the polynomial.

The constants that precede the variables are called **coefficients**.

The highest value of the exponent of x in a polynomial is called the **degree** of the polynomial. So, $-4x^3 + x$ has a degree of 3, while $-2x^5 + x^3 + 4x + 1$ has a degree of 5. When multiplying polynomials, the degree of the result will be the sum of the degrees of the two polynomials being multiplied.

To add polynomials, add the coefficients of like powers of x. For example, $(-2x^5 + x^3 + 4x + 1) + (-4x^3 + x) = -2x^5 + (1 - 4)x^3 + (4 + 1)x + 1 = -2x^5 - 3x^3 + 5x + 1$.

Likewise, subtraction of polynomials is performed by subtracting coefficients of like powers of x. So, $(-2x^5 + x^3 + 4x + 1) - (-4x^3 + x) = -2x^5 + (1 + 4)x^3 + (4 - 1)x + 1 = -2x^5 + 5x^3 + 3x + 1$.

To multiply two polynomials, multiply each term of the first polynomial by each term of the second polynomial and add the results. For example, $(4x^2 + x)(-x^3 + x) = 4x^2(-x^3) + 4x^2(x) + x(-x^3) + x(x) = -4x^5 + 4x^3 - x^4 + x^2$. In the case where each polynomial has two terms, like in this example, some students find it helpful to remember this as multiplying the First terms, then the Outer terms, then the Inner terms, and finally the Last terms, with the mnemonic **FOIL**. For longer polynomials, the multiplication process is the same, but there will be, of course, more terms, and there is no common mnemonic to remember each combination.

The process of **factoring** a polynomial means to write the polynomial as a product of other (generally simpler) polynomials. Here is an example: $x^2 - 4x + 3 = (x - 1)(x - 3)$. If a certain monomial divides every term of the polynomial, factor it out of each term, for example:

$$4x^3 + 16x^2 = 4x^2(x + 4).$$

$$x^2 + 2xy + y^2 = (x + y)^2 \text{ or } x^2 - 2xy + y^2 = (x - y)^2$$

$$x^2 - y^2 = (x + y)(x - y)$$

$$x^3 + y^3 = (x + y)(x^2 - xy + y^2)$$

$$x^3 - y^3 = (x - y)(x^2 + xy + y^2)$$

$$x^3 + 3x^2y + 3xy^2 + y^3 = (x + y)^3 \text{ and } x^3 - 3x^2y + 3xy^2 - y^3 = (x - y)^3$$

It sometimes can be necessary to rewrite the polynomial in some clever way before applying the above rules. Consider the problem of factoring $x^4 - 1$. This does not immediately look like any of the cases for which there are rules. However, it's possible to think of this polynomial as $x^4 - 1 = (x^2)^2 - (1^2)^2$, and now apply the third rule in the above list to simplify this: $(x^2)^2 - (1^2)^2 = (x^2 + 1^2)(x^2 - 1^2) = (x^2 + 1)(x^2 - 1)$.

Zeros of Polynomials

Finding the zeros of polynomial equations are the points at which the graph of the equation crosses the x-axis. Factors can be used to find the zeros of a polynomial equation or function. The degree of the equation shows the number of possible zeros. If the highest exponent on the independent variable is 4, then the degree is 4, and the number of possible zeros is 4. If there are complex solutions, the number of roots is less than the degree.

Given the equation $y = x^2 + 7x + 6$, y can be set equal to zero, and the polynomial can be factored. The equation turns into $0 = (x + 1)(x + 6)$, where $x = -1$ and $x = -6$ are the zeros. Since this is a quadratic equation, the shape of the graph will be a parabola. Knowing that zeros represent the points where the parabola crosses the x-axis, the maximum or minimum point is the only other piece needed to sketch a rough graph of the function. By looking at the function in standard form, the coefficient of x

is positive; therefore, the parabola opens *up*. Using the zeros and the minimum, the following rough sketch of the graph can be constructed:

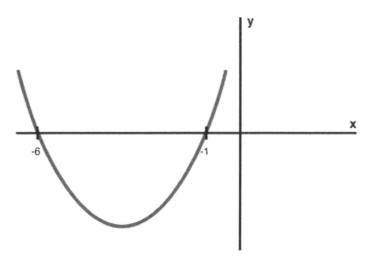

Polynomial Identities

Difference of squares refers to a binomial composed of the difference of two squares. For example, $a^2 - b^2$ is a difference of squares. It can be written $(a)^2 - (b)^2$, and it can be factored into $(a - b)(a + b)$. Recognizing the difference of squares allows the expression to be rewritten easily because of the form it takes. For some expressions, factoring consists of more than one step. When factoring, it's important to always check to make sure that the result cannot be factored further. If it can, then the expression should be split further. If it cannot be, the factoring step is complete, and the expression is completely factored.

A sum and difference of cubes is another way to factor a polynomial expression. When the polynomial takes the form of addition or subtraction of two terms that can be written as a cube, a formula is given. The following graphic shows the factorization of a difference of cubes:

$$a^3 - b^3 = (a - b)(a^2 + ab + b^2)$$

same sign
opposite sign
always +

This form of factoring can be useful in finding the zeros of a function of degree 3. For example, when solving $x^3 - 27 = 0$, this rule needs to be used. $x^3 - 27$ is first written as the difference two cubes, $(x)^3 - (3)^3$ and then factored into $(x - 3)(x^2 + 3x + 9)$. This expression may not be factored any further. Each factor is then set equal to zero. Therefore, one solution is found to be $x = 3$, and the other two solutions must be found using the quadratic formula. A sum of squares would have a similar process. The formula for factoring a sum of squares is $a^3 + b^3 = (a + b)(a^2 - ab + b^2)$.

The opposite of factoring is multiplying. Multiplying a square of a binomial involves the following rules: $(a + b)^2 = a^2 + 2ab + b^2$ and $(a - b)^2 = a^2 - 2ab + b^2$. The binomial theorem for expansion can be

135

used when the exponent on a binomial is larger than 2, and the multiplication would take a long time. The binomial theorem is given as:

$$(a+b)^n = \sum_{k=0}^{n} \binom{n}{k} a^{n-k} b^k \qquad \text{where} \quad \binom{n}{k} = \frac{n!}{k!(n-k)!}.$$

The **Remainder Theorem** can be helpful when evaluating polynomial functions $P(x)$ for a given value of x. A polynomial can be divided by $(x-a)$, if there is a remainder of 0. This also means that $P(a) = 0$ and $(x-a)$ is a factor of $P(x)$. In a similar sense, if P is evaluated at any other number b, $P(b)$ is equal to the remainder of dividing $P(x)$ by $(x-b)$.

Exponential and Logarithmic Equations

Exponential Functions or Equations

An **exponential function** is a function of the form $f(x) = b^x$, where b is a positive real number other than 1. In such a function, b is called the **base**. An exponential equation takes the same form, but is typically written as $y = b^x$.

The **domain** of an exponential function is all real numbers, and the *range* is all positive real numbers. There will always be a horizontal asymptote of $y = 0$ on one side. If b is greater than 1, then the graph will be increasing moving to the right. If b is less than 1, then the graph will be decreasing moving to the right. Exponential functions are one-to-one. The basic exponential function graph will go through the point (0,1).

Example: Solve $5^{x+1} = 25$.

Get the x out of the exponent by rewriting the equation $5^{x+1} = 5^2$ so that both sides have a base of 5.

Since the bases are the same, the exponents must be equal to each other.

This leaves $x + 1 = 2$ or $x = 1$.

To check the answer, the x-value of 1 can be substituted back into the original equation.

Logarithmic Functions and Equations

A **logarithmic function** is an inverse for an exponential function. The inverse of the base b exponential function is written as $\log_b(x)$, and is called the *base b* **logarithm**. The domain of a logarithm is all positive real numbers. It has the properties that $\log_b(b^x) = x$. For positive real values of x, $b^{\log_b(x)} = x$.

When there is no chance of confusion, the parentheses are sometimes skipped for logarithmic functions: $\log_b(x)$ may be written as $\log_b x$. For the special number e, the base e logarithm is called the **natural logarithm** and is written as $\ln x$. Logarithms are one-to-one.

When working with logarithmic functions, it is important to remember the following properties. Each one can be derived from the definition of the logarithm as the inverse to an exponential function:

$$\log_b 1 = 0$$

$$\log_b b = 1$$

$$\log_b b^p = p$$

$$\log_b MN = \log_b M + \log_b N$$

$$\log_b \frac{M}{N} = \log_b M - \log_b N$$

$$\log_b M^p = p \log_b M$$

When solving equations involving exponentials and logarithms, the following fact should be used:

If f is a one-to-one function, $a = b$ is equivalent to $f(a) = f(b)$.

Using this, together with the fact that logarithms and exponentials are inverses, allows manipulations of the equations to isolate the variable.

Example: Solve $4 = \ln(x - 4)$.

Using the definition of a logarithm, the equation can be changed to $e^4 = e^{\ln(x-4)}$.

The functions on the right side cancel with a result of $e^4 = x - 4$.

This then gives $x = 4 + e^4$.

Trigonometry

Trigonometric Functions

Trigonometric functions are built out of two basic functions, the sine and cosine, written as $\sin \theta$ and $\cos \theta$ respectively. Note that similar to logarithms, it is customary to drop the parentheses as long as the result is not confusing.

The sine and cosine are defined using the **unit circle**. If θ is the angle going counterclockwise around the origin from the *x*-axis, then the point on the unit circle in that direction will have the coordinates $(\cos \theta, \sin \theta)$.

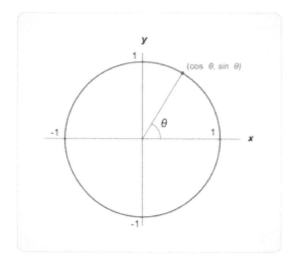

Since the angle returns to the start every 2π radians (or 360 degrees), the graph of these functions will be **periodic**, with period 2π. This means that the graph repeats itself as one moves along the *x*-axis because $\sin \theta = \sin(\theta + 2\pi)$. Cosine is works similarly.

From the unit circle definition, the **sine** function starts at 0 when $\theta = 0$. It grows to 1 as θ grows to $\pi/2$, and then back to 0 at $\theta = \pi$. Then it decreases to -1 as θ grows to $3\pi/2$, and back up to 0 at $\theta = 2\pi$.

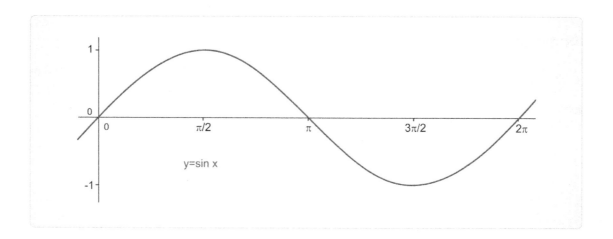

The graph of the **cosine** is similar. The cosine will start at 1, decreasing to 0 at $\pi/2$ and continuing to decrease to -1 at $\theta = \pi$. Then, it grows to 0 as θ grows to $3\pi/2$ and back up to 1 at $\theta = 2\pi$.

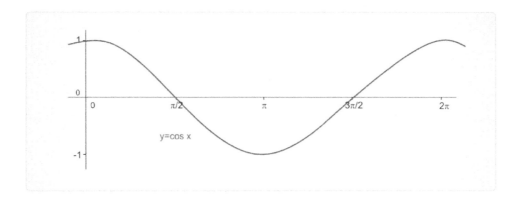

Another trigonometric function, which is frequently used, is the **tangent** function. This is defined as the following equation: $\tan \theta = \frac{\sin \theta}{\cos \theta}$.

The tangent function is a period of π rather than 2π because the sine and cosine functions have the same absolute values after a change in the angle of π, but flip their signs. Since the tangent is a ratio of the two functions, the changes in signs cancel.

The tangent function will be zero when the sine is zero, and it will have a vertical asymptote whenever cosine is zero. The following graph shows the tangent function:

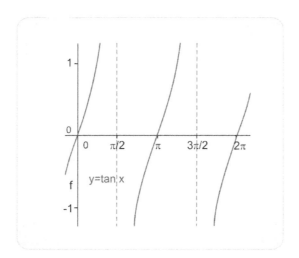

Three other trigonometric functions are sometimes useful. These are the *reciprocal trigonometric functions*, so named because they are just the reciprocals of sine, cosine, and tangent. They are the **cosecant**, defined as $\csc \theta = \frac{1}{\sin \theta}$, the **secant**, $\sec \theta = \frac{1}{\cos \theta}$, and the **cotangent**, $\cot \theta = \frac{1}{\tan \theta}$. Note that from the definition of tangent, $\cot \theta = \frac{\cos \theta}{\sin \theta}$.

In addition, there are three identities that relate the trigonometric functions to one another:

$$\cos\theta = \sin(\frac{\pi}{2} - \theta)$$

$$\csc\theta = \sec\left(\frac{\pi}{2} - \theta\right)$$

$$\cot\theta = \tan(\frac{\pi}{2} - \theta)$$

Here is a list of commonly-needed values for trigonometric functions, given in radians, for the first quadrant:

Table for trigonometric functions

$\sin 0 = 0$	$\cos 0 = 1$	$\tan 0 = 0$
$\sin\frac{\pi}{6} = \frac{1}{2}$	$\cos\frac{\pi}{6} = \frac{\sqrt{3}}{2}$	$\tan\frac{\pi}{6} = \frac{\sqrt{3}}{3}$
$\sin\frac{\pi}{4} = \frac{\sqrt{2}}{2}$	$\cos\frac{\pi}{4} = \frac{\sqrt{2}}{2}$	$\tan\frac{\pi}{4} = 1$
$\sin\frac{\pi}{3} = \frac{\sqrt{3}}{2}$	$\cos\frac{\pi}{3} = \frac{1}{2}$	$\tan\frac{\pi}{3} = \sqrt{3}$
$\sin\frac{\pi}{2} = 1$	$\cos\frac{\pi}{2} = 0$	$\tan\frac{\pi}{2} = undefined$
$\csc 0 = undefined$	$\sec 0 = 1$	$\cot 0 = undefined$
$\csc\frac{\pi}{6} = 2$	$\sec\frac{\pi}{6} = \frac{2\sqrt{3}}{3}$	$\cot\frac{\pi}{6} = \sqrt{3}$
$\csc\frac{\pi}{4} = \sqrt{2}$	$\sec\frac{\pi}{4} = \sqrt{2}$	$\cot\frac{\pi}{4} = 1$
$\csc\frac{\pi}{3} = \frac{2\sqrt{3}}{3}$	$\sec\frac{\pi}{3} = 2$	$\cot\frac{\pi}{3} = \frac{\sqrt{3}}{3}$
$\csc\frac{\pi}{2} = 1$	$\sec\frac{\pi}{2} = undefined$	$\cot\frac{\pi}{2} = 0$

To find the trigonometric values in other quadrants, complementary angles can be used. The **complementary angle** is the smallest angle between the *x*-axis and the given angle.

Once the complementary angle is known, the following rule is used:

For an angle θ with complementary angle x, the absolute value of a trigonometry function evaluated at θ is the same as the absolute value when evaluated at x.

The correct sign is used based on the functions sine and cosine are given by the x and y coordinates on the unit circle.

- Sine will be positive in quadrants I and II and negative in quadrants III and IV.

- Cosine will be positive in quadrants I and IV, and negative in II and III.

- Tangent will be positive in I and III, and negative in II and IV.

- The signs of the reciprocal functions will be the same as the sign of the function of which they are a reciprocal.

Example: Find $\sin \frac{3\pi}{4}$.

First, the complementary angle must be found.

This angle is in the II quadrant, and the angle between it and the x-axis is $\frac{\pi}{4}$.

Now, $\sin \frac{\pi}{4} = \frac{\sqrt{2}}{2}$.

Since this is in the II quadrant, sine takes on positive values (the y coordinate is positive in the II quadrant).

Therefore, $\sin \frac{3\pi}{4} = \frac{\sqrt{2}}{2}$.

In addition to the six trigonometric functions defined above, there are inverses for these functions. However, since the trigonometric functions are not one-to-one, one can only construct inverses for them on a restricted domain.

Usually, the domain chosen will be $[0, \pi)$ for cosine and $(-\frac{\pi}{2}, \frac{\pi}{2}]$ for sine. The inverse for tangent can use either of these domains. The inverse functions for the trigonometric functions are also called *arc functions*. In addition to being written with a -1 in the exponent to denote that the function is an inverse, they will sometimes be written with an "a" or "arc" in front of the function name, so $\cos^{-1} \theta = a\cos \theta = \arccos \theta$.

When solving equations that involve trigonometric functions, there are often multiple solutions. For example, $2 \sin \theta = \sqrt{2}$ can be simplified to $\sin \theta = \frac{\sqrt{2}}{2}$. This has solutions $\theta = \frac{\pi}{4}, \frac{3\pi}{4}$, but in addition, because of the periodicity, any integer multiple of 2π can also be added to these solutions to find another solution.

The full set of solutions is $\theta = \frac{\pi}{4} + 2\pi k, \frac{3\pi}{4} + 2\pi k$ for all integer values of k. It is very important to remember to find all possible solutions when dealing with equations that involve trigonometric functions.

The name *trigonometric* comes from the fact that these functions play an important role in the geometry of triangles, particularly right triangles.

Consider the right triangle shown in this figure:

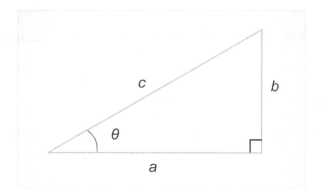

The following hold true:

- $c \sin \theta = b$

- $c \cos \theta = a$

- $\tan \theta = \frac{b}{a}$

- $b \csc \theta = c$

- $a \sec \theta = c$

- $\cot \theta = \frac{a}{b}$

Remember also the angles of a triangle must add up to π radians (180 degrees).

Solving Trigonometric Functions

Solving trigonometric functions can be done with a knowledge of the unit circle and the trigonometric identities. It requires the use of opposite operations combined with trigonometric ratios for special triangles. For example, the problem may require solving the equation $2 \cos^2 x - \sqrt{3} \cos x = 0$ for the values of x between 0 and 180 degrees. The first step is to factor out the $\cos x$ term, resulting in $\cos x \, (2 \cos x - \sqrt{3}) = 0$. By the factoring method of solving, each factor can be set equal to zero: $\cos x = 0$ and $(2 \cos x - \sqrt{3}) = 0$. The second equation can be solved to yield the following equation: $\cos x = \frac{\sqrt{3}}{2}$. Now that the value of x is found, the trigonometric ratios can be used to find the solutions of $x = 30$ and 90 degrees.

Solving trigonometric functions requires the use of algebra to isolate the variable and a knowledge of trigonometric ratios to find the value of the variable. The unit circle can be used to find answers for special triangles. Beyond those triangles, a calculator can be used to solve for variables within the trigonometric functions.

Using Graphing Calculators to Solve Trigonometric Problems

In addition to algebraic techniques, problems involving trigonometric functions can be solved using graphing calculators. For example, given an equation, both sides of the equals sign first need to be graphed as separate equations in the same window on the calculator. The point(s) of intersection are then found by zooming into an appropriate window. The points of intersection are the solutions. For example, consider $\cos x = \frac{1}{2}$. Solving this equation involves graphing both $y = \cos x$ and $y = \frac{1}{2}$ in the same window. If the calculator is set in Radians mode, the screen zoomed in from $[0, 2\pi]$ on the x-axis would look like:

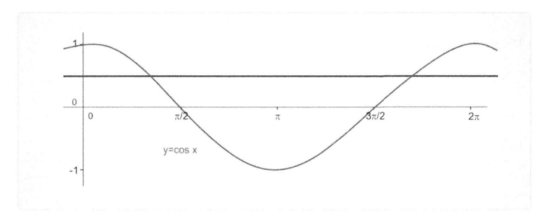

The trace function on the calculator allows the user to zoom into the point of intersection to obtain the two solutions from $[0, 2\pi]$. The periodic nature of the function must then be taken into consideration to obtain the entire solution set, which contains an infinite number of solutions.

Practice Questions

1. What is the product of the following expression?

$$(4x - 8)(5x^2 + x + 6)$$

    a. $20x^3 - 36x^2 + 16x - 48$
    b. $6x^3 - 41x^2 + 12x + 15$
    c. $204 + 11x^2 - 37x - 12$
    d. $2x^3 - 11x^2 - 32x + 20$

2. What is the y-intercept for $y = x^2 + 3x - 4$?
    a. $y = 1$
    b. $y = -4$
    c. $y = 3$
    d. $y = 4$

3. What is the value of b in the equation: $5b - 4 = 2b + 17$?
    a. 13
    b. 24
    c. 7
    d. 21

4. If $-3(x + 4) \geq x + 8$, what is the value of x?
    a. $x = 4$
    b. $x \geq 2$
    c. $x \geq -5$
    d. $x \leq -5$

5. Karen gets paid a weekly salary and a commission for every sale that she makes. The table below shows the number of sales and her pay for different weeks.

Sales	2	7	4	8
Pay	$380	$580	$460	$620

Which of the following equations represents Karen's weekly pay?
    a. $y = 90x + 200$
    b. $y = 90x - 200$
    c. $y = 40x + 300$
    d. $y = 40x - 300$

Answer Explanations

1. A: Finding the product means distributing one polynomial over the other so that each term in the first is multiplied by each term in the second. Then, like terms can be collected. Multiplying the factors yields the expression:

$$20x^3 + 4x^2 + 24x - 40x^2 - 8x - 48$$

Collecting like terms means adding the x^2 terms and adding the x terms. The final answer after simplifying the expression is:

$$20x^3 - 36x^2 + 16x - 48$$

By making this change in position over time into a rate, the speed becomes ten meters in two seconds or five meters in one second.

2. B: The y-intercept of an equation is found where the x-value is zero. Plugging zero into the equation for x, the first two terms cancel out, leaving -4.

3. C: To solve for the value of b, both sides of the equation need to be equalized.

Start by cancelling out the lower value of -4 by adding 4 to both sides:

$$5b - 4 = 2b + 17$$

$$5b = 2b + 21$$

The variable b is the same on each side, so subtract the lower 2b from each side:

$$5b = 2b + 21$$

$$3b = 21$$

Then divide both sides by 3 to get the value of b:

$$\frac{3b}{3} = \frac{21}{3}$$

$$b = 7$$

4. D: $x \leq -5$. When solving a linear equation or inequality:

Distribution is performed if necessary:

$$-3(x + 4) \rightarrow -3x - 12 \geq x + 8$$

This means that any like terms on the same side of the equation/inequality are combined.

The equation/inequality is manipulated to get the variable on one side. In this case, subtracting x from both sides produces:

$$-4x - 12 \geq 8$$

The variable is isolated using inverse operations to undo addition/subtraction. Adding 12 to both sides produces $-4x \geq 20$.

The variable is isolated using inverse operations to undo multiplication/division. Remember if dividing by a negative number, the relationship of the inequality reverses, so the sign is flipped. In this case, dividing by -4 on both sides produces $x \leq -5$.

5. C: $y = 40x + 300$

In this scenario, the variables are the number of sales and Karen's weekly pay. The weekly pay depends on the number of sales. Therefore, weekly pay is the dependent variable (y), and the number of sales is the independent variable (x). Each pair of values from the table can be written as an ordered pair (x, y): (2, 380), (7, 580), (4, 460), (8, 620).

The ordered pairs can be substituted into the equations to see which creates true statements (both sides equal) for each pair. Even if one ordered pair produces equal values for a given equation, the other three ordered pairs must be checked. The only equation which is true for all four ordered pairs is

$$y = 40x + 300$$

$$380 = 40(2) + 300 \rightarrow 380 = 380$$

$$580 = 40(7) + 300 \rightarrow 580 = 580$$

$$460 = 40(4) + 300 \rightarrow 460 = 460$$

$$620 = 40(8) + 300 \rightarrow 620 = 620$$

Information and Ideas

Identifying the Main Idea

Topics and main ideas are critical parts of any writing. The **topic** is the subject matter of the piece, and it is a broader, more general term. The **main idea** is what the writer wants to say about that topic. The topic can be expressed in a word or two, but the main idea should be a complete thought.

The topic and main idea are usually easy to recognize in nonfiction writing. An author will likely identify the topic immediately in the first sentence of a passage or essay. The main idea is also typically presented in the introductory paragraph of an essay. In a single passage, the main idea may be identified in the first or last sentence, but will likely be directly stated and easily recognized by the reader. Because it is not always stated immediately in a passage, it's important to carefully read the entire passage to identify the main idea.

Also remember that when most authors write, they want to make a point or send a message. This point or message of a text is known as the theme. Authors may state themes explicitly, like in *Aesop's Fables*. More often, especially in modern literature, readers must infer the theme based on text details. Usually after carefully reading and analyzing an entire text, the theme emerges. Typically, the longer the piece, the more themes you will encounter, though often one theme dominates the rest, as evidenced by the author's purposeful revisiting of it throughout the passage.

The main idea should not be confused with the thesis statement. A thesis statement is a clear statement of the writer's specific stance, and can often be found in the introduction of a nonfiction piece. The main idea is more of an overview of the entire piece, while the thesis is a specific sentence found in that piece.

In order to illustrate the main idea, a writer will use **supporting details** in a passage. These details can provide evidence or examples to help make a point. Supporting details are most commonly found in nonfiction pieces that seek to inform or persuade the reader.

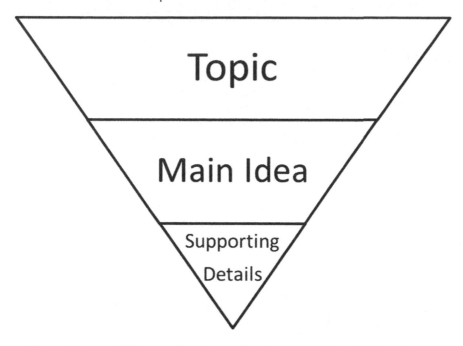

As a reader, you will want to carefully examine the author's supporting details to be sure they are credible. Consider whether they provide evidence of the author's point and whether they directly support the main idea. You might find that an author has used a shocking statistic to grab your attention, but that the statistic doesn't really support the main idea, so it isn't being effectively used in the piece.

Identifying Supporting Details

Supporting details help readers better develop and understand the main idea. Supporting details answer questions like *who, what, where, when, why,* and *how.* Different types of supporting details include examples, facts and statistics, anecdotes, and sensory details.

Persuasive and informative texts often use supporting details. In persuasive texts, authors attempt to make readers agree with their points of view, and supporting details are often used as "selling points." If authors make a statement, they need to support the statement with evidence in order to adequately persuade readers. Informative texts use supporting details such as examples and facts to inform readers. Review the previous "Cheetahs" passage to find examples of supporting details.

Cheetahs

Cheetahs are one of the fastest mammals on the land, reaching up to 70 miles an hour over short distances. Even though cheetahs can run as fast as 70 miles an hour, they usually only have to run half that speed to catch up with their choice of prey. Cheetahs cannot maintain a fast pace over long periods of time because their bodies will overheat. After a chase, cheetahs need to rest for approximately 30 minutes prior to eating or returning to any other activity.

In the example, supporting details include:

- Cheetahs reach up to 70 miles per hour over short distances.
- They usually only have to run half that speed to catch up with their prey.
- Cheetahs will overheat if they exert a high speed over longer distances.
- Cheetahs need to rest for 30 minutes after a chase.

Look at the diagram below (applying the cheetah example) to help determine the hierarchy of topic, main idea, and supporting details.

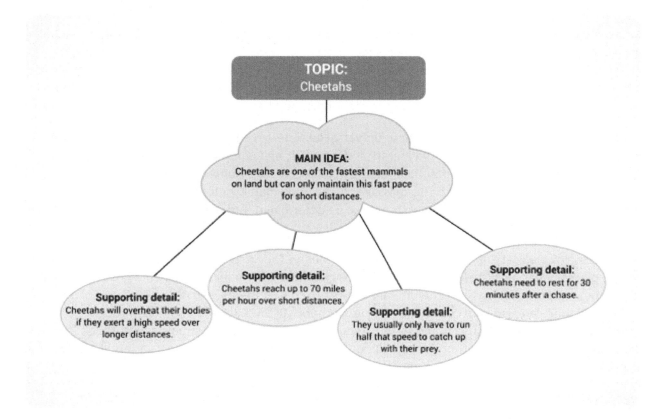

Identifying Central Ideas or Themes of a Text

The **theme** is the central message of a fictional work, whether that work is structured as prose, drama, or poetry. It is the heart of what an author is trying to say to readers through the writing, and theme is largely conveyed through literary elements and techniques.

In literature, a theme can often be determined by considering the over-arching narrative conflict within the work. Though there are several types of conflicts and several potential themes within them, the following are the most common:

A. Individual against the self—relevant to themes of self-awareness, internal struggles, pride, coming of age, facing reality, fate, free will, vanity, loss of innocence, loneliness, isolation, fulfillment, failure, and disillusionment

B. Individual against nature—relevant to themes of knowledge vs. ignorance, nature as beauty, quest for discovery, self-preservation, chaos and order, circle of life, death, and destruction of beauty

C. Individual against society—relevant to themes of power, beauty, good, evil, war, class struggle, totalitarianism, role of men/women, wealth, corruption, change vs. tradition, capitalism, destruction, heroism, injustice, and racism

D. Individual against another individual—relevant to themes of hope, loss of love or hope, sacrifice, power, revenge, betrayal, and honor

For example, in Hawthorne's *The Scarlet Letter*, one possible narrative conflict could be the individual against the self, with a relevant theme of internal struggles. This theme is alluded to through characterization—Dimmesdale's moral struggle with his love for Hester and Hester's internal struggles with the truth and her daughter, Pearl. It's also alluded to through plot—Dimmesdale's suicide and Hester helping the very townspeople who initially condemned her.

Sometimes, a text can convey a **message** or **universal lesson**—a truth or insight that the reader infers from the text, based on analysis of the literary and/or poetic elements. This message is often presented as a statement. For example, a potential message in Shakespeare's *Hamlet* could be "Revenge is what ultimately drives the human soul." This message can be immediately determined through plot and characterization in numerous ways, but it can also be determined through the setting of Norway, which is bordering on war.

Authors employ a variety of techniques to present a theme. They may compare or contrast characters, events, places, ideas, or historical or invented settings to speak thematically. They may use analogies, metaphors, similes, allusions, or other literary devices to convey the theme. An author's use of diction, syntax, and tone can also help convey the theme. Authors will often develop themes through the development of characters, use of the setting, repetition of ideas, use of symbols, and through contrasting value systems. Authors of both fiction and nonfiction genres will use a variety of these techniques to develop one or more themes.

Regardless of the literary genre, there are commonalities in how authors, playwrights, and poets develop themes or central ideas.

Authors often do research, the results of which contribute to theme. In prose fiction and drama, this research may include real historical information about the setting the author has chosen or include elements that make fictional characters, settings, and plots seem realistic to the reader. In nonfiction, research is critical since the information contained within this literature must be accurate and, moreover, accurately represented.

In fiction, authors present a narrative conflict that will contribute to the overall theme. In fiction, this conflict may involve the storyline itself and some trouble within characters that needs resolution. In nonfiction, this conflict may be an explanation or commentary on factual people and events.

Authors will sometimes use character motivation to convey theme, such as in the example from *Hamlet* regarding revenge. In fiction, the characters an author creates will think, speak, and act in ways that effectively convey the theme to readers. In nonfiction, the characters are factual, as in a biography, but authors pay particular attention to presenting those motivations to make them clear to readers.

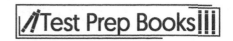

Authors also use literary devices as a means of conveying theme. For example, the use of moon symbolism in Mary Shelley's *Frankenstein* is significant as its phases can be compared to the phases that the Creature undergoes as he struggles with his identity.

The selected point of view can also contribute to a work's theme. The use of first-person point of view in a fiction or non-fiction work engages the reader's response differently than third person point of view. The central idea or theme from a first-person narrative may differ from a third-person limited text.

In literary nonfiction, authors usually identify the purpose of their writing, which differs from fiction, where the general purpose is to entertain. The purpose of nonfiction is usually to inform, persuade, or entertain the audience. The stated purpose of a non-fiction text will drive how the central message or theme, if applicable, is presented.

Authors identify an audience for their writing, which is critical in shaping the theme of the work. For example, the audience for J.K. Rowling's *Harry Potter* series would be different than the audience for a biography of George Washington. The audience an author chooses to address is closely tied to the purpose of the work. The choice of an audience also drives the choice of language and level of diction an author uses. Ultimately, the intended audience determines the level to which that subject matter is presented and the complexity of the theme.

Central Ideas in Informational Texts

Informational text is specifically designed to relate factual information, and although it is open to a reader's interpretation and application of the facts, the structure of the presentation is carefully designed to lead the reader to a particular conclusion or central idea. When reading informational text, it is important that readers are able to understand its organizational structure as the structure often directly relates to an author's intent to inform and/or persuade the reader.

The first step in identifying the text's structure is to determine the thesis or main idea. The thesis statement and organization of a work are closely intertwined. A **thesis statement** indicates the writer's purpose and may include the scope and direction of the text. It may be presented at the beginning of a text or at the end, and it may be explicit or implicit.

Once a reader has a grasp of the thesis or main idea of the text, he or she can better determine its organizational structure. Test takers are advised to read informational text passages more than once in order to comprehend the material fully. It is also helpful to examine any text features present in the text including the table of contents, index, glossary, headings, footnotes, and visuals. The analysis of these features and the information presented within them can offer additional clues about the central idea and structure of a text.

The following questions should be asked when considering structure:

- How does the author assemble the parts to make an effective whole argument?

- Is the passage linear in nature and if so, what is the timeline or thread of logic?

- What is the presented order of events, facts, or arguments? Are these effective in contributing to the author's thesis?

- How can the passage be divided into sections? How are they related to each other and to the main idea or thesis?

- What key terms are used to indicate the organization?

Next, test takers should skim the passage, noting the first line or two of each body paragraph—the **topic sentences**—and the conclusion. Key **transitional terms**, such as *on the other hand*, *also*, *because*, *however*, *therefore*, *most importantly*, and *first*, within the text can also signal organizational structure. Based on these clues, readers should then be able to identify what type of organizational structure is being used. The following organizational structures are most common:

A. **Problem/solution**: organized by an analysis/overview of a problem, followed by potential solution(s)

B. **Cause/effect**: organized by the effects resulting from a cause or the cause(s) of a particular effect

C. **Spatial order**: organized by points that suggest location or direction—e.g., top to bottom, right to left, outside to inside

D. **Chronological/sequence order**: organized by points presented to indicate a passage of time or through purposeful steps/stages

E. **Comparison/contrast**: organized by points that indicate similarities and/or differences between two things or concepts

F. **Order of importance**: organized by priority of points, often most significant to least significant or vice versa

Summaries of Main Points and Supporting Details

Creating an outline that identifies the **main ideas** of a passage as well as the **supporting details** is a helpful tool in effectively summarizing a text. Most outlines will include a title that reveals the topic of the text and is usually a single phrase or word, such as "whales." If the passage is divided up into paragraphs, or the paragraphs into sections, each paragraph or section will have its own main idea. These "main ideas" are usually depicted in outlines as roman numerals. Next, writers use supporting details in order to support or prove the main ideas. The supporting details should be listed underneath each main idea in the outline.

For example:

Title: Whales
I. Killer whales
       a. Highly social
       b. Apex predator
II. Humpback whales
       a. Males produce "song"
       b. Targeted for whaling industry
III. Beluga whales
       a. Complex sense of hearing
       b. Slow swimmers

Making an outline is a useful method of summarization because it forces the reader to deconstruct the text as a whole and to identify only the most important parts of the text.

Ideas from a text can also be organized using graphic organizers. A **graphic organizer** is a way to simplify information and take key points from the text. A graphic organizer such as a timeline may have an event listed for a corresponding date on the timeline while an outline may have an event listed under a key point that occurs in the text. Each reader needs to create the type of graphic organizer that works the best for him or her in terms of being able to recall information from a story. Examples include a **spider-map**, which takes a main idea from the story and places it in a bubble with supporting points branching off the main idea. An **outline** is useful for diagramming the main and supporting points of the entire story, and a **Venn diagram** classifies information as separate or overlapping.

Writing a summary is similar to creating an outline. In both instances, the reader wants to relay the most important parts of the text without being too verbose. A **summary** of a text should begin with stating the main idea of that text. Then, the reader must decide which supporting details are absolutely essential to the main idea of the text and leave any irrelevant information out of the summary. A summary shouldn't be too brief—readers should include important details depicted in the text—but it also shouldn't be too long either. The appeal of a summary to an audience is that they are able to receive the message of the text without being distracted by the style or detours of the author.

Another effective reading comprehension strategy is paraphrasing. Paraphrasing is usually longer than a summary. **Paraphrasing** is taking the author's text and rewriting, or "translating," it into their own words. A tip for paraphrasing is to read a passage over three times. Once you read the passage and understand what the author is saying, cover the original passage and begin to write everything you remember from that passage into your own words. Usually, if you understand the content well enough, you will have translated the main idea of the author into your own words with your own writing style. An effective paraphrase will be as long as the original passage but will have a different writing structure.

Summaries of Informational Texts

Informational text is written material that has the primary function of imparting information about a topic. It is often written by someone with expertise in the topic and directed at an audience that has less knowledge of the topic. Informational texts are written in a different fashion from storytelling or narrative texts. Typically, this type of text has several organizational and structural differences from narrative text. In informational texts, there are features such as charts, graphs, photographs, headings and subheadings, glossaries, indexes, bibliographies, or other guidance features. With the aid of technology, embedded hyperlinks and video content are also sometimes included. Informational material may be written to compare and contrast, be explanatory, link cause and effect, provide opinion, persuade the reader, or serve a number of other purposes. Finally, informational texts typically use a different style of language than narrative texts, which instead, focus more on storytelling.

Making Logical Inferences

Critical readers should be able to make inferences. Making an **inference** requires the reader to read between the lines and look for what is *implied* rather than what is directly stated. That is, using information that *is* known from the text, the reader is able to make a logical assumption about information that is *not* directly stated but is probably true. Read the following passage:

"Hey, do you wanna meet my new puppy?" Jonathan asked.

"Oh, I'm sorry but please don't—" Jacinta began to protest, but before she could finish Jonathan had already opened the passenger side door of his car and a perfect white ball of fur came bouncing towards Jacinta.

"Isn't he the cutest?" beamed Jonathan.

"Yes—achoo!—he's pretty—aaaachooo!!—adora—aaa—aaaachoo!" Jacinta managed to say in between sneezes. "But if you don't mind, I—I—achoo!—need to go inside."

Which of the following can be inferred from Jacinta's reaction to the puppy?
   a. she hates animals
   b. she is allergic to dogs
   c. she prefers cats to dogs
   d. she is angry at Jonathan

An inference requires the reader to consider the information presented and then form their own idea about what is probably true. Based on the details in the passage, what is the best answer to the question? Important details to pay attention to include the tone of Jacinta's dialogue, which is overall polite and apologetic, as well as her reaction itself, which is a long string of sneezes. Answer Choices *A* and *D* both express strong emotions ("hates" and "angry") that are not evident in Jacinta's speech or actions. Answer Choice *C* mentions cats, but there is nothing in the passage to indicate Jacinta's feelings about cats. Answer Choice *B*, "she is allergic to dogs," is the most logical choice—based on the fact that she began sneezing as soon as a fluffy dog approached her, it makes sense to guess that Jacinta might be allergic to dogs. So even though Jacinta never directly states, "Sorry, I'm allergic to dogs!" using the clues in the passage, it is still reasonable to guess that this is true.

Making inferences is crucial for readers of literature, because literary texts often avoid presenting complete and direct information to readers about characters' thoughts or feelings, or they present this information in an unclear way, leaving it up to the reader to interpret clues given in the text. In order to make inferences while reading, readers should ask themselves:

   • What details are being presented in the text?
   • Is there any important information that seems to be missing?
   • Based on the information that the author *does* include, what else is probably true?
   • Is this inference reasonable based on what is already known?

Critical Thinking Skills

It's important to read any piece of writing critically. The goal is to discover the point and purpose of what the author is writing about through analysis. It's also crucial to establish the point or stance the author has taken on the topic of the piece. After determining the author's perspective, readers can then more effectively develop their own viewpoints on the subject of the piece.

It is important to distinguish between fact and opinion when reading a piece of writing. A **fact** is information that can be proven true. If information can be disproved, it is not a fact. For example, water freezes at or below thirty-two degrees Fahrenheit. An argument stating that water freezes at seventy degrees Fahrenheit cannot be supported by data and is therefore not a fact. Facts tend to be associated with science, mathematics, and statistics. **Opinions** are information open to debate. Opinions are often tied to subjective concepts like equality, morals, and rights. They can also be controversial.

Authors often use words like *think, feel, believe,* or *in my opinion* when expressing opinion, but these words won't always appear in an opinion piece, especially if it is formally written. An author's opinion may be backed up by facts, which gives it more credibility, but that opinion should not be taken as fact.

A critical reader should be suspect of an author's opinion, especially if it is only supported by other opinions.

Fact	Opinion
There are 9 innings in a game of baseball.	Baseball games run too long.
James Garfield was assassinated on July 2, 1881.	James Garfield was a good president.
McDonalds has stores in 118 countries.	McDonalds has the best hamburgers.

Critical readers examine the facts used to support an author's argument. They check the facts against other sources to be sure those facts are correct. They also check the validity of the sources used to be sure those sources are credible, academic, and/or peer- reviewed. Consider that when an author uses another person's opinion to support his or her argument, even if it is an expert's opinion, it is still only an opinion and should not be taken as fact. A strong argument uses valid, measurable facts to support ideas. Even then, the reader may disagree with the argument as it may be rooted in his or her personal beliefs.

An authoritative argument may use the facts to sway the reader. In the example of global warming, many experts differ in their opinions of what alternative fuels can be used to aid in offsetting it. Because of this, a writer may choose to only use the information and expert opinion that supports his or her viewpoint.

If the argument is that wind energy is the best solution, the author will use facts that support this idea. That same author may leave out relevant facts on solar energy. The way the author uses facts can influence the reader, so it's important to consider the facts being used, how those facts are being presented, and what information might be left out.

Critical readers should also look for errors in the argument such as logical fallacies and bias. A **logical fallacy** is a flaw in the logic used to make the argument. Logical fallacies include slippery slope, straw man, and begging the question. Authors can also reflect **bias** if they ignore an opposing viewpoint or present their side in an unbalanced way. A strong argument considers the opposition and finds a way to refute it. Critical readers should look for an unfair or one-sided presentation of the argument and be skeptical, as a bias may be present. Even if this bias is unintentional, if it exists in the writing, the reader should be wary of the validity of the argument.

Readers should also look for the use of **stereotypes**, which refer to specific groups. Stereotypes are often negative connotations about a person or place and should always be avoided. When a critical reader finds stereotypes in a piece of writing, they should immediately be critical of the argument and consider the validity of anything the author presents. Stereotypes reveal a flaw in the writer's thinking and may suggest a lack of knowledge or understanding about the subject.

Rhetoric

The ACCUPLACER will test a reader's ability to identify an author's use of rhetoric within text passages. **Rhetoric** is the use of positional or persuasive language to convey one or more central ideas. The idea behind the use of rhetoric is to convince the reader of something. Its use is meant to persuade or motivate the reader. An author may choose to appeal to their audience through logic, emotion, the use of ideology, or by conveying that the central idea is timely, and thus, important to the reader. There are a variety of rhetorical techniques an author can use to achieve this goal.

An author may choose to use traditional elements of style to persuade the reader. They may also use a story's setting, mood, characters, or a central conflict to build emotion in the reader. Similarly, an author may choose to use specific techniques such as alliteration, irony, metaphor, simile, hyperbole, allegory, imagery, onomatopoeia, and personification to persuasively illustrate one or more central ideas they wish the reader to adopt. In order to be successful in a standardized reading comprehension test situation, a reader needs to be well acquainted in recognizing rhetoric and rhetorical devices.

Determining How Ideas or Details Inform the Author's Argument

Once a reader has determined an author's thesis or main idea, he or she will need to understand how textual evidence supports interpretation of that thesis or main idea. Test takers will be asked direct questions regarding an author's main idea and may be asked to identify evidence that would support those ideas. This will require test takers to comprehend literal and figurative meanings within the text passage, be able to draw inferences from provided information, and be able to separate important evidence from minor supporting details. It's often helpful to skim test questions and answer options prior to critically reading informational text; however, test takers should avoid the temptation to solely look for the correct answers. Just trying to find the "right answer" may cause test takers to miss important supporting textual evidence. Making mental note of test questions is only helpful as a guide when reading.

After identifying an author's thesis or main idea, a test taker should look at the supporting details that the author provides to back up his or her assertions, identifying those additional pieces of information that help expand the thesis. From there, test takers should examine the additional information and related details for credibility, the author's use of outside sources, and be able to point to direct evidence that supports the author's claims. It's also imperative that test takers be able to identify what is strong support and what is merely additional information that is nice to know but not necessary. Being able to make this differentiation will help test takers effectively answer questions regarding an author's use of supporting evidence within informational text.

Identifying Elements of Style

A writer's style is unique. The combinations of elements are carefully designed to create an effect on the reader. For example, the novels of J.K. Rowling are very different in style than the novels of Stephen King, yet both are designed to tell a compelling tale and to entertain readers. Furthermore, the articles found in *National Geographic* are vastly different from those a reader may encounter in *People* magazine, yet both have the same objective: to inform the reader. The difference is in the elements of style.

While there are many elements of style an author can employ, it's important to look at three things: the words they choose to use, the voice an author selects, and the fluency of sentence structure. Word choice is critical in persuasive or pictorial writing. While effective authors will choose words that are succinct, different authors will choose various words based on what they are trying to accomplish. For example, a reader would not expect to encounter the same words in a gothic novel that they would read in a scholastic article on gene therapy. An author whose intent is to paint a picture of a foreboding scene, will choose different words than an author who wants to persuade the reader that a particular political party has the most sound, ideological platform. A romance novelist will sound very different than a true crime writer.

The voice an author selects is also important to note. An author's voice is that element of style that indicates their personality. It's important that authors move us as readers; therefore, they will choose a voice that helps them do that. An author's voice may be satirical or authoritative. It may be light-hearted or serious in tone. It may be silly or humorous as well. Voice, as an element of style, can be vague in nature and difficult to identify, since it's also referred to as an author's tone, but it is that element unique to the author. It is the author's "self." A reader can expect an author's voice to vary across literary genres. A non-fiction author will generally employ a more neutral voice than an author of fiction, but use caution when trying to identify voice. Do not confuse an author's voice with a particular character's voice.

Another critical element of style involves how an author structures their sentences. An effective writer—one who wants to paint a vivid picture or strongly illustrate a central idea—will use a variety of sentence structures and sentence lengths. A reader is more likely to be confused if an author uses choppy, unrelated sentences. Similarly, a reader will become bored and lose interest if an author repeatedly uses the same sentence structure. Good writing is fluent. It flows. Varying sentence structure keeps a reader engaged and helps reading comprehension. Consider the following example:

> The morning started off early. It was bright out. It was just daylight. The moon was still in the sky. He was tired from his sleepless night.

Then consider this text:

> Morning hit hard. He didn't remember the last time light hurt this bad. Sleep had been absent, and the very thought of moving towards the new day seemed like a hurdle he couldn't overcome.

Note the variety in sentence structure. The second passage is more interesting to read because the sentence fluency is more effective. Both passages paint the picture of a central character's reaction to dawn, but the second passage is more effective because it uses a variety of sentences and is more fluent than the first.

Elements of style can also include more recognizable components such as a story's setting, the type of narrative an author chooses, the mood they set, and the character conflicts employed. The ability to effectively understand the use of rhetoric demands the reader take note of an author's word choices, writing voice, and the ease of fluency employed to persuade, entertain, illustrate, or otherwise captivate a reader.

Tone

An author's **tone** is the use of particular words, phrases, and writing style to convey an overall meaning. Tone expresses the author's attitude towards a particular topic. For example, a historical reading passage may begin like the following:

> The presidential election of 1960 ushered in a new era, a new Camelot, a new phase of forward thinking in U.S. politics that embraced brash action, unrest, and responded with admirable leadership.

From this opening statement, a reader can draw some conclusions about the author's attitude towards President John F. Kennedy. Furthermore, the reader can make additional, educated guesses about the state of the Union during the 1960 presidential election. By close reading, the test taker can determine that the repeated use of the word *new* and words such as *admirable leadership* indicate the author's tone of admiration regarding the President's boldness. In addition, the author assesses that the era during President Kennedy's administration was problematic through the use of the words *brash action* and *unrest.* Therefore, if a test taker encountered a test question asking about the author's use of tone and their assessment of the Kennedy administration, the test taker should be able to identify an answer indicating admiration. Similarly, if asked about the state of the Union during the 1960s, a test taker should be able to correctly identify an answer indicating political unrest.

When identifying an author's tone, the following list of words may be helpful. This is not an inclusive list. Generally, parts of speech that indicate attitude will also indicate tone:

- Comical
- Angry
- Ambivalent
- Scary
- Lyrical
- Matter-of-fact
- Judgmental
- Sarcastic
- Malicious
- Objective
- Pessimistic
- Patronizing
- Gloomy
- Instructional
- Satirical
- Formal
- Casual

Message

An author's **message** is the same as the overall meaning of a passage. It is the main idea, or the main concept the author wishes to convey. An author's message may be stated outright, or it may be implied. Regardless, the test taker will need to use careful reading skills to identify an author's message or purpose.

Often, the message of a particular passage can be determined by thinking about why the author wrote the information. Many historical passages are written to inform and to teach readers established, factual information. However, many historical works are also written to convey biased ideas to readers. Gleaning bias from an author's message in a historical passage can be difficult, especially if the reader is presented with a variety of established facts as well. Readers tend to accept historical writing as factual. This is not always the case. Any discerning reader who has tackled historical information on topics such as United States political party agendas can attest that two or more works on the same topic may have completely different messages supporting or refuting the value of the identical policies. Therefore, it is important to critically assess an author's message separate from factual information. One author, for example, may point to the rise of unorthodox political candidates in an election year based on the failures of the political party in office while another may point to the rise of the same candidates in the same election year based on the current party's successes. The historical facts of what has occurred leading up to an election year are not in refute. Labeling those facts as a failure or a success is a bias within an author's overall message, as is excluding factual information in order to further a particular point. In a standardized testing situation, a reader must be able to critically assess what the author is trying to say separate from the historical facts that surround their message.

Using the example of Lincoln's Gettysburg Address, a test question may ask the following:

> What is the message the author is trying to convey through this address?

Then they will ask the test taker to select an answer that best expresses Lincoln's message to his audience. Based on the options given, a test taker should be able to select the answer expressing the idea that Lincoln's audience should recognize the efforts of those who died in the war as a sacrifice to preserving human equality and self-government.

Effect

The **effect** an author wants to convey is when an author wants to impart a particular mood in their message. An author may want to challenge a reader's intellect, inspire imagination, or spur emotion. An author may present information to appeal to a physical, aesthetic, or transformational sense.

Take the following text as an example:

> In 1963, Martin Luther King stated "I have a dream." The gathering at the Lincoln Memorial was the beginning of the Civil Rights movement and, with its reference to the Emancipation Proclamation, Dr. King's words electrified those who wanted freedom and equality while rising from hatred and slavery. It was the beginning of radical change.

The test taker may be asked about the effect this statement might have on King's audience. Through careful reading of the passage, the test taker should be able to choose an answer that best identifies an effect of grabbing the audience's attention. The historical facts are in place: King made the speech in 1963 at the Lincoln Memorial, kicked off the civil rights movement, and referenced the Emancipation Proclamation. The words *electrified* and *radical change* indicate the effect the author wants the reader to understand as a result of King's speech. In this historical passage, facts are facts. However, the author's message goes above the facts to indicate the effect the message had on the audience and, in addition, the effect the event should have on the reader.

How an Author's Word Choice Conveys Attitude and Shapes Meaning, Style, and Tone

Authors choose their words carefully in order to artfully depict meaning, style, and tone, which is most commonly inferred through the use of adjectives and verbs. The **tone** is the predominant emotion present in the text and represents the attitude or feelings that an author has towards a character or event.

To review, an **adjective** is a word used to describe something, and usually precedes the **noun**, a person, place, or object. A **verb** is a word describing an action. For example, the sentence "The scary woodpecker ate the spider" includes the adjective "scary," the noun "woodpecker," and the verb "ate." Reading this sentence may rouse some negative feelings, as the word "scary" carries a negative charge. The **charge** is the emotional connotation that can be derived from the adjectives and verbs and is either positive or negative. Recognizing the charge of a particular sentence or passage is an effective way to understand the meaning and tone the author is trying to convey.

Many authors have conflicting charges within the same text, but a definitive tone can be inferred by understanding the meaning of the charges relative to each other. It's important to recognize key **conjunctions**, or words that link sentences or clauses together. There are several types and subtypes of conjunctions. Three are most important for reading comprehension:

- **Cumulative conjunctions** add one statement to another.
    - Examples: *and, both, also, as well as, not only*
    - e.g. The juice is sweet *and* sour.
- **Adversative conjunctions** are used to contrast two clauses.
    - Examples: *but, while, still, yet, nevertheless*
    - e.g. She was tired, *but* she was happy.
- **Alternative conjunctions** express two alternatives.
    - Examples: *or, either, neither, nor, else, otherwise*
    - e.g. He must eat, *or* he will die.

Identifying the meaning and tone of a text can be accomplished with the following techniques:

- Identify the adjectives and verbs.
- Recognize any important conjunctions.
- Label the adjectives and verbs as positive or negative.
- Understand what the charge means about the text.

160

To demonstrate these steps, examine the following passage from the classic children's poem, "The Sheep":

> Lazy sheep, pray tell me why
>
> In the pleasant fields you lie,
>
> Eating grass, and daisies white,
>
> From the morning till the night?
>
> Everything can something do,
>
> But what kind of use are you?

> –Taylor, Jane and Ann. "The Sheep."

This selection is a good example of conflicting charges that work together to express an overall tone. Following the first two steps, identify the adjectives, verbs, and conjunctions within the passage. For this example, the adjectives are <u>underlined</u>, the verbs are in **bold**, and the conjunctions *italicized*:

> <u>Lazy</u> sheep, pray **tell** me why
>
> In the <u>pleasant</u> fields you **lie**,
>
> **Eating** grass, and daisies <u>white,</u>
>
> From the morning till the night?
>
> Everything can something do,
>
> *But* what kind of use are you?

For step three, read the passage and judge whether feelings of positivity or negativity arose. Then assign a charge to each of the words that were outlined. This can be done in a table format, or simply by writing a + or − next to the word.

The word <u>lazy</u> carries a negative connotation; it usually denotes somebody unwilling to work. To **tell** someone something has an exclusively neutral connotation, as it depends on what's being told, which has not yet been revealed at this point, so a charge can be assigned later. The word <u>pleasant</u> is an inherently positive word. To **lie** could be positive or negative depending on the context, but as the subject (the sheep) is lying in a pleasant field, then this is a positive experience. **Eating** is also generally positive.

After labeling the charges for each word, it might be inferred that the tone of this poem is happy and maybe even admiring or innocuously envious. However, notice the adversative conjunction, "but" and what follows. The author has listed all the pleasant things this sheep gets to do all day, but the tone changes when the author asks, "What kind of use are you?" Asking someone to prove their value is a rather hurtful thing to do, as it implies that the person asking the question doesn't believe the subject has any value, so this could be listed under negative charges. Referring back to the verb **tell**, after

reading the whole passage, it can be deduced that the author is asking the sheep to tell what use the sheep is, so this has a negative charge.

+	−
• Pleasant • Lie in fields • From morning to night	• Lazy • Tell me • What kind of use are you

Upon examining the charges, it might seem like there's an even amount of positive and negative emotion in this selection, and that's where the conjunction "but" becomes crucial to identifying the tone. The conjunction "but" indicates there's a contrasting view to the pleasantness of the sheep's daily life, and this view is that the sheep is lazy and useless, which is also indicated by the first line, "lazy sheep, pray tell me why."

It might be helpful to look at questions pertaining to tone. For this selection, consider the following question:

The author of the poem regards the sheep with a feeling of what?
a. Respect
b. Disgust
c. Apprehension
d. Intrigue

Considering the author views the sheep as lazy with nothing to offer, Choice *A* appears to reflect the opposite of what the author is feeling.

Choice *B* seems to mirror the author's feelings towards the sheep, as laziness is considered a disreputable trait, and people (or personified animals, in this case) with unfavorable traits might be viewed with disgust.

Choice *C* doesn't make sense within context, as laziness isn't usually feared.

Choice *D* is tricky, as it may be tempting to argue that the author is intrigued with the sheep because they ask, "pray tell me why." This is another out-of-scope answer choice as it doesn't *quite* describe the feelings the author experiences and there's also a much better fit in Choice *B*.

An author's choice of words—also referred to as **diction**—helps to convey meaning in a particular way. Through diction, an author can convey a particular tone—e.g., a humorous tone, a serious tone—in order to support the thesis in a meaningful way to the reader.

Connotation and Denotation

Connotation is when an author chooses words or phrases that invoke ideas or feelings other than their literal meaning. An example of the use of connotation is the word *cheap*, which suggests something is poor in value or negatively describes a person as reluctant to spend money. When something or someone is described this way, the reader is more inclined to have a particular image or feeling. Thus, connotation can be a very effective language tool in creating emotion and swaying opinion. However, connotations are sometimes hard to pin down because varying emotions can be associated with a word. Generally, though, connotative meanings tend to be fairly consistent within a specific cultural group.

Denotation refers to words or phrases that mean exactly what they say. It is helpful when a writer wants to present hard facts or vocabulary terms with which readers may be unfamiliar. Some examples of denotation are the words *inexpensive* and *frugal*. *Inexpensive* refers to the cost of something, not its value, and *frugal* indicates that a person is conscientiously watching his or her spending. These terms do not elicit the same emotions that *cheap* does.

Authors sometimes choose to use both, but what they choose and when they use it is what critical readers need to differentiate. One method isn't inherently better than the other; however, one may create a better effect, depending upon an author's intent. If, for example, an author's purpose is to inform, to instruct, and to familiarize readers with a difficult subject, his or her use of connotation may be helpful. However, it may also undermine credibility and confuse readers. An author who wants to create a credible, scholarly effect in his or her text would most likely use denotation, which emphasizes literal, factual meaning and examples.

Technical Language

Test takers and critical readers alike should be very aware of technical language used within informational text. **Technical language** refers to terminology that is specific to a particular industry and is best understood by those specializing in that industry. This language is fairly easy to differentiate since it will most likely be unfamiliar to readers. It's critical to be able to define technical language either by the author's written definition, through the use of an included glossary—if offered—or through context clues that help readers clarify word meaning.

Analyzing the Structure of a Text

Good writing is not merely a random collection of sentences. No matter how well written, sentences must relate and coordinate appropriately with one another. If not, the writing seems random, haphazard, and disorganized. Therefore, good writing must be organized, where each sentence fits a larger context and relates to the sentences around it.

Transition Words

The writer should act as a guide, showing the reader how all the sentences fit together. Consider this seat belt example:

> Seat belts save more lives than any other automobile safety feature. Many studies show that airbags save lives as well. Not all cars have airbags. Many older cars don't. Air bags aren't entirely reliable. Studies show that in 15% of accidents, airbags don't deploy as designed. Seat belt malfunctions are extremely rare.

There's nothing wrong with any of these sentences individually, but together they're disjointed and difficult to follow. The best way for the writer to communicate information is through the use of transition words. Here are examples of transition words and phrases that tie sentences together, enabling a more natural flow:

To show causality: *as a result, therefore*, and *consequently*
To compare and contrast: *however, but,* and *on the other hand*
To introduce examples: *for instance, namely,* and *including*
To show order of importance: *foremost, primarily, secondly,* and *lastly*

Note that this is not a complete list of transitions. There are many more that can be used; however, most fit into these or similar categories. The important point is that the words should clearly show the relationship between sentences, supporting information, and the main idea.

Here is an update to the previous example using transition words. These changes make it easier to read and bring clarity to the writer's points:

> Seat belts save more lives than any other automobile safety feature. Many studies show that airbags save lives as well; however, not all cars have airbags. For instance, some older cars don't. Furthermore, air bags aren't entirely reliable. For example, studies show that in 15% of accidents, airbags don't deploy as designed, but, on the other hand, seat belt malfunctions are extremely rare.

Also, test takers should be prepared to analyze whether the writer is using the best transition word or phrase for the situation. For example, the sentence: "As a result, seat belt malfunctions are extremely rare" doesn't make sense in the context above because the writer is trying to show the contrast between seat belts and airbags, not the causality.

Logical Sequence

Even if the writer includes plenty of information to support their point, the writing is only coherent when the information is in a logical order. First, the writer should introduce the main idea, whether for a paragraph, a section, or the entire piece. Second, they should present evidence to support the main idea by using transitional language. This shows the reader how the information relates to the main idea and to the sentences around it. The writer should then take time to interpret the information, making sure necessary connections are obvious to the reader. Finally, the writer can summarize the information in a closing section.

Though most writing follows this pattern, it isn't a set rule. Sometimes writers change the order for effect. For example, the writer can begin with a surprising piece of supporting information to grab the reader's attention, and then transition to the main idea. Thus, if a passage doesn't follow the logical order, don't immediately assume it's wrong. However, most writing usually settles into a logical sequence after a nontraditional beginning.

Introductions and Conclusions

Examining the writer's strategies for introductions and conclusions puts the reader in the right mindset to interpret the rest of the text. Look for methods the writer might use for introductions such as:

> Stating the main point immediately, followed by outlining how the rest of the piece supports this claim.

> Establishing important, smaller pieces of the main idea first, and then grouping these points into a case for the main idea.

> Opening with a quotation, anecdote, question, seeming paradox, or other piece of interesting information, and then using it to lead to the main point.

Whatever method the writer chooses, the introduction should make their intention clear, establish their voice as a credible one, and encourage a person to continue reading.

Conclusions tend to follow a similar pattern. In them, the writer restates their main idea a final time, often after summarizing the smaller pieces of that idea. If the introduction uses a quote or anecdote to grab the reader's attention, the conclusion often makes reference to it again. Whatever way the writer chooses to arrange the conclusion, the final restatement of the main idea should be clear and simple for the reader to interpret. Finally, conclusions shouldn't introduce any new information.

Organization

Depending on what the author is attempting to accomplish, certain formats or text structures work better than others. For example, a sequence structure might work for narration but not when identifying similarities and differences between dissimilar concepts. Similarly, a comparison-contrast structure is not useful for narration. It's the author's job to put the right information in the correct format.

Readers should be familiar with the five main literary structures:

1. **Sequence** structure (sometimes referred to as the order structure) is when the order of events proceed in a predictable order. In many cases, this means the text goes through the plot elements: exposition, rising action, climax, falling action, and resolution. Readers are introduced to characters, setting, and conflict in the exposition. In the rising action, there's an increase in tension and suspense. The climax is the height of tension and the point of no return. Tension decreases during the falling action. In the resolution, any conflicts presented in the exposition are solved, and the story concludes. An informative text that is structured sequentially will often go in order from one step to the next.

2. In the **problem-solution** structure, authors identify a potential problem and suggest a solution. This form of writing is usually divided into two paragraphs and can be found in informational texts. For example, cell phone, cable, and satellite providers use this structure in manuals to help customers troubleshoot or identify problems with services or products.

3. When authors want to discuss similarities and differences between separate concepts, they arrange thoughts in a **comparison-contrast** paragraph structure. Venn diagrams are an effective graphic organizer for comparison-contrast structures because they feature two overlapping circles that can be used to organize similarities and differences. A comparison-contrast essay organizes one paragraph based on similarities and another based on differences. A comparison-contrast essay can also be arranged with the similarities and differences of individual traits addressed within individual paragraphs. Words such as *however, but*, and *nevertheless* help signal a contrast in ideas.

4. **Descriptive** writing structure is designed to appeal to your senses. Much like an artist who constructs a painting, good descriptive writing builds an image in the reader's mind by appealing to the five senses: sight, hearing, taste, touch, and smell. However, overly descriptive writing can become tedious; sparse descriptions can make settings and characters seem flat. Good authors strike a balance by applying descriptions only to passages, characters, and settings that are integral to the plot.

5. Passages that use the **cause and effect** structure are simply asking *why* by demonstrating some type of connection between ideas. Words such as *if, since, because, then*, or *consequently* indicate relationship. By switching the order of a complex sentence, the writer can rearrange the emphasis on different clauses. Saying *If Sheryl is late, we'll miss the dance* is different from saying *We'll miss the dance if Sheryl is late*. One emphasizes Sheryl's tardiness while the other emphasizes missing the dance. Paragraphs can also be arranged in a cause and effect format. Since the format—before and after—is

165

sequential, it is useful when authors wish to discuss the impact of choices. Researchers often apply this paragraph structure to the scientific method.

Point of View

Point of view is an important writing device to consider. In fiction writing, point of view refers to who tells the story or from whose perspective readers are observing as they read. In non-fiction writing, the **point of view** refers to whether authors refer to themselves, their readers, or choose not to refer to either. Whether fiction or nonfiction, the author will carefully consider the impact the perspective will have on the purpose and main point of the writing.

G. **First-person point of view**: The story is told from the writer's perspective. In fiction, this would mean that the main character is also the narrator. First-person point of view is easily recognized by the use of personal pronouns such as *I, me, we, us, our, my,* and *myself.*

H. **Third-person point of view**: In a more formal essay, this would be an appropriate perspective because the focus should be on the subject matter, not the writer or the reader. Third-person point of view is recognized using the pronouns *he, she, they,* and *it*. In fiction writing, third person point of view has a few variations.

I. **Third-person limited point of view**: Refers to a story told by a narrator who has access to the thoughts and feelings of just one character.

J. **Third-person omniscient point of view**: The narrator has access to the thoughts and feelings of all the characters.

Point of View	Pronouns Used
First person	I, me, we, us, our, my, myself
Second person	You, your, yourself
Third person	He, she, it, they

K. **Third-person objective point of view**: The narrator is like a fly on the wall and can see and hear what the characters do and say but does not have access to their thoughts and feelings.

L. **Second-person point of view**: This point of view isn't commonly used in fiction or non-fiction writing because it directly addresses the reader using the pronouns *you, your,* and *yourself.* Second-person perspective is more appropriate in direct communication, such as business letters or emails.

How the Position and Purpose Shape the Text

When it comes to authors' writings, readers should always identify a position or stance. No matter how objective a piece may seem, assume the author has preconceived beliefs. Reduce the likelihood of accepting an invalid argument by looking for multiple articles on the topic, including those with varying opinions. If several opinions point in the same direction, and are backed by reputable peer-reviewed sources, it's more likely the author has a valid argument. Positions that run contrary to widely held beliefs and existing data should invite scrutiny. There are exceptions to the rule, so be a careful consumer of information.

Though themes, symbols, and motifs are buried deep within the text and can sometimes be difficult to infer, an author's purpose is usually obvious from the beginning. There are four purposes of writing: to inform, to persuade, to describe, and to entertain. Informative writings present facts in an accessible way and are also known as expository writing. Persuasive writing appeals to emotions and logic to inspire the reader to adopt a specific stance. Be wary of this type of writing, as it often lacks objectivity. Descriptive writing is designed to paint a picture in the reader's mind, while writings that entertain are often narratives designed to engage and delight the reader.

The various writing styles are usually blended, with one purpose dominating the rest. For example, a persuasive piece might begin with a humorous tale to make readers more receptive to the persuasive message, or a recipe in a cookbook designed to inform might be preceded by an entertaining anecdote that makes the recipe more appealing.

Determining an Author's Point of View

A **rhetorical strategy**—also referred to as a **rhetorical mode**—is the structural way an author chooses to present his/her argument. Though the terms noted below are similar to the organizational structures noted earlier, these strategies do not imply that the entire text follows the approach. For example, a cause and effect organizational structure is solely that, nothing more. A persuasive text may use cause and effect as a strategy to convey a singular point. Thus, an argument may include several of the strategies as the author strives to convince his or her audience to take action or accept a different point of view. It's important that readers are able to identify an author's thesis and position on the topic in order to be able to identify the careful construction through which the author speaks to the reader. The following are some of the more common rhetorical strategies:

- **Cause and effect**—establishing a logical correlation or causation between two ideas
- **Classification/division**—the grouping of similar items together or division of something into parts
- **Comparison/contrast**—the distinguishing of similarities/differences to expand on an idea
- **Definition**—used to clarify abstract ideas, unfamiliar concepts, or to distinguish one idea from another
- **Description**—use of vivid imagery, active verbs, and clear adjectives to explain ideas
- **Exemplification**—the use of examples to explain an idea
- **Narration**—anecdotes or personal experience to present or expand on a concept
- **Problem/solution**—presentation of a problem or problems, followed by proposed solution(s)

Identifying Rhetorical Devices

If a writer feels strongly about a subject, or has a passion for it, strong words and phrases can be chosen. Think of the types of rhetoric (or language) our politicians use. Each word, phrase, and idea is carefully crafted to elicit a response. Hopefully, that response is one of agreement to a certain point of view, especially among voters. Authors use the same types of language to achieve the same results. For example, the word "bad" has a certain connotation, but the words "horrid," "repugnant," and "abhorrent" paint a far better picture for the reader. They're more precise. They're interesting to read and they should all illicit stronger feelings in the reader than the word "bad." An author generally uses other devices beyond mere word choice to persuade, convince, entertain, or otherwise engage a reader.

Rhetorical devices are those elements an author utilizes in painting sensory, and hopefully persuasive ideas to which a reader can relate. They are numerable. Test takers will likely encounter one or more standardized test questions addressing various rhetorical devices. This study guide will address the more common types: alliteration, irony, metaphor, simile, hyperbole, allegory, imagery, onomatopoeia, and personification, providing examples of each.

Alliteration is a device that uses repetitive beginning sounds in words to appeal to the reader. Classic tongue twisters are a great example of alliteration. *She sells sea shells down by the sea shore* is an extreme example of alliteration. Authors will use alliterative devices to capture a reader's attention. It's interesting to note that marketing also utilizes alliteration in the same way. A reader will likely remember products that have the brand name and item starting with the same letter. Similarly, many songs, poems, and catchy phrases use this device. It's memorable. Use of alliteration draws a reader's attention to ideas that an author wants to highlight.

Irony is a device that authors use when pitting two contrasting items or ideas against each other in order to create an effect. It's frequently used when an author wants to employ humor or convey a sarcastic tone. Additionally, it's often used in fictional works to build tension between characters, or between a particular character and the reader. An author may use *verbal irony* (sarcasm), *situational irony* (where actions or events have the opposite effect than what's expected), and *dramatic irony* (where the reader knows something a character does not). Examples of irony include:

- **Dramatic Irony**: An author describing the presence of a hidden killer in a murder mystery, unbeknownst to the characters but known to the reader.

- **Situational Irony**: An author relating the tale of a fire captain who loses her home in a five-alarm conflagration.

- **Verbal Irony**: This is where an author or character says one thing but means another. For example, telling a police officer "Thanks a lot" after receiving a ticket.

Metaphor is a device that uses a figure of speech to paint a visual picture of something that is not literally applicable. Authors relate strong images to readers, and evoke similar strong feelings using metaphors. Most often, authors will mention one thing in comparison to another more familiar to the reader. It's important to note that metaphors do not use the comparative words "like" or "as." At times, metaphors encompass common phrases such as clichés. At other times, authors may use mixed metaphors in making identification between two dissimilar things. Examples of metaphors include:

- An author describing a character's anger as *a flaming sheet of fire*.
- An author relating a politician as having been a folding chair under close questioning.
- A novel's character telling another character to *take a flying hike*.
- Shakespeare's assertion that *all the world's a stage*.

Simile is a device that compares two dissimilar things using the words "like" and "as." When using similes, an author tries to catch a reader's attention and use comparison of unlike items to make a point. Similes are commonly used and often develop into figures of speech and catch phrases.

Examples of similes include:

- An author describing a character as having a complexion like a faded lily.

- An investigative journalist describing his interview subject as being like cold steel and with a demeanor hard as ice.

- An author asserting the current political arena is just like a three-ring circus and as dry as day old bread.

Similes and metaphors can be confusing. When utilizing simile, an author will state one thing is like another. A metaphor states one thing is another. An example of the difference would be if an author states a character is *just like a fierce tiger and twice as angry,* as opposed to stating the character *is a fierce tiger and twice as angry.*

Hyperbole is simply an exaggeration that is not taken literally. A potential test taker will have heard or employed hyperbole in daily speech, as it is a common device we all use. Authors will use hyperbole to draw a reader's eye toward important points and to illicit strong emotional and relatable responses. Examples of hyperbole include:

- An author describing a character as being as big as a house and twice the circumference of a city block.

- An author stating the city's water problem as being old as the hills and more expensive than a king's ransom in spent tax dollars.

- A journalist stating the mayoral candidate died of embarrassment when her tax records were made public.

Allegories are stories or poems with hidden meanings, usually a political or moral one. Authors will frequently use allegory when leading the reader to a conclusion. Allegories are similar to parables, symbols, and analogies. Often, an author will employ the use of allegory to make political, historical, moral, or social observations. As an example, Jonathan Swift's work *Gulliver's Travels into Several Remote Nations of the World* is an allegory in and of itself. The work is a political allegory of England during Jonathan Swift's lifetime. Set in the travel journal style plot of a giant amongst smaller people, and a smaller Gulliver amongst the larger, it is a commentary on Swift's political stance of existing issues of his age. Many fictional works are entire allegories in and of themselves. George Orwell's *Animal Farm* is a story of animals that conquer man and form their own farm society with swine at the top; however, it is not a literal story in any sense. It's Orwell's political allegory of Russian society during and after the Communist revolution of 1917. Other examples of allegory in popular culture include:

- Aesop's fable "The Tortoise and the Hare," which teaches readers that being steady is more important than being fast and impulsive.

- The popular *Hunger Games* by Suzanne Collins that teaches readers that media can numb society to what is truly real and important.

- Dr. Seuss's *Yertle the Turtle* which is a warning against totalitarianism and, at the time it was written, against the despotic rule of Adolf Hitler.

Imagery is a rhetorical device that an author employs when they use visual or descriptive language to evoke a reader's emotion. Use of imagery as a rhetorical device is broader in scope than this study guide addresses, but in general, the function of imagery is to create a vibrant scene in the reader's imagination and, in turn, tease the reader's ability to identify through strong emotion and sensory experience. In the simplest of terms, imagery, as a rhetoric device, beautifies literature.

An example of poetic imagery is below:

Pain has an element of blank

It cannot recollect

When it began, or if there were

A day when it was not.

It has no future but itself,

Its infinite realms contain

Its past, enlightened to perceive

New periods of pain.

In the above poem, Emily Dickenson uses strong imagery. Pain is equivalent to an "element of blank" or of nothingness. Pain cannot recollect a beginning or end, as if it was a person (see *personification* below). Dickenson appeals to the reader's sense of a painful experience by discussing the unlikelihood that discomfort sees a future, but does visualize a past and present. She simply indicates that pain, through the use of imagery, is cyclical and never ending. Dickenson's theme is one of painful depression, and it is through the use of imagery that she conveys this to her readers.

Onomatopoeia is the author's use of words that create sound. Words like *pop* and *sizzle* are examples of onomatopoeia. When an author wants to draw a reader's attention in an auditory sense, they will use onomatopoeia. An author may also use onomatopoeia to create sounds as interjection or commentary. Examples include:

- An author describing a cat's vocalization as the kitten's chirrup echoed throughout the empty cabin.
- A description of a campfire as crackling and whining against its burning green wood.
- An author relating the sound of a car accident as *metallic screeching against crunching asphalt*.
- A description of an animal roadblock as being *a symphonic melody of groans, baas, and moans*.

Personification is a rhetorical device that an author uses to attribute human qualities to inanimate objects or animals. Once again, this device is useful when an author wants the reader to strongly relate to an idea. As in the example of George Orwell's *Animal Farm*, many of the animals are given the human abilities to speak, reason, apply logic, and otherwise interact as humans do. This helps the reader see how easily it is for any society to segregate into the haves and the have-nots through the manipulation of power. Personification is a device that enables the reader to empathize through human experience.

Examples of personification include:

- An author describing the wind as *whispering through the trees*.

- A description of a stone wall as being a hardened, unmovable creature made of cement and brick.

- An author attributing a city building as having slit eyes and an unapproachable, foreboding façade.

- An author describing spring as a beautiful bride, blooming in white, ready for summer's matrimony.

When identifying rhetorical devices, look for words and phrases that capture one's attention. Make note of the author's use of comparison between the inanimate and the animate. Consider words that make the reader feel sounds and envision imagery. Pay attention to the rhythm of fluid sentences and to the use of words that evoke emotion. The ability to identify rhetorical devices is another step in achieving successful reading comprehension and in being able to correctly answer standardized questions related to those devices.

Understanding Methods Used to Appeal to a Specific Audience

In an argument or persuasive text, an author will strive to sway readers to an opinion or conclusion. To be effective, an author must consider his or her intended audience. Although an author may write text for a general audience, he or she will use methods of appeal or persuasion to convince that audience. Aristotle asserted that there were three methods or modes by which a person could be persuaded. These are referred to as **rhetorical appeals**.

The three main types of rhetorical appeals are shown in the following graphic.

Ethos, also referred to as an **ethical appeal**, is an appeal to the audience's perception of the writer as credible (or not), based on their examination of their ethics and who the writer is, his/her experience or incorporation of relevant information, or his/her argument. For example, authors may present testimonials to bolster their arguments. The reader who critically examines the veracity of the testimonials and the credibility of those giving the testimony will be able to determine if the author's use of testimony is valid to his or her argument. In turn, this will help the reader determine if the author's thesis is valid. An author's careful and appropriate use of technical language can create an overall knowledgeable effect and, in turn, act as a convincing vehicle when it comes to credibility. Overuse of technical language, however, may create confusion in readers and obscure an author's overall intent.

Pathos, also referred to as **emotional appeal**, is an appeal to the audience's sense of identity, self-interest, or emotions. A critical reader will notice when the author is appealing to pathos through anecdotes and descriptions that elicit an emotion such as anger or pity. Readers should also beware of factual information that uses generalization to appeal to the emotions. While it's tempting to believe an author is the source of truth in his or her text, an author who presents factual information as universally true, consistent throughout time, and common to all groups is using **generalization**. Authors who exclusively use generalizations without specific facts and credible sourcing are attempting to sway readers solely through emotion.

Logos, also referred to as a **logical appeal**, is an appeal to the audience's ability to see and understand the logic in a claim offered by the writer. A critical reader has to be able to evaluate an author's arguments for validity of reasoning and for sufficiency when it comes to argument.

Synthesis

Synthesis

Synthesis in reading involves the ability to fully comprehend text passages, and then going further by making new connections to see things in a new or different way. It involves a full thought process and requires readers to change the way they think about what they read. The ACCUPLACER will require a test taker to integrate new information that he or she already knows, and demonstrate an ability to express new thoughts.

Synthesis goes further than summary. When **summarizing**, a reader collects all of the information an author presents in a text passage, and restates it in an effective manner. Synthesis requires that the test taker not only summarize reading material, but be able to express new ideas based on the author's message. It is a full culmination of all reading comprehension strategies. It will require the test taker to order, recount, summarize, and recreate information into a whole new idea.

In utilizing synthesis, a reader must be able to form mental images about what they read, recall any background information they have about the topic, ask critical questions about the material, determine the importance of points an author makes, make inferences based on the reading, and finally be able to form new ideas based on all of the above skills. Synthesis requires the reader to make connections, visualize concepts, determine their importance, ask questions, make inferences, then fully synthesize all of this information into new thought.

Making Connections in Reading
There are three helpful thinking strategies to keep in mind when attempting to synthesize text passages:

- Think about how the content of a passage relates to life experience.
- Think about how the content of a passage relates to other text.
- Think about how the content of a passage relates to the world in general.

When reading a given passage, the test taker should actively think about how the content relates to their life experience. While the author's message may express an opinion different from what the reader believes, or express ideas with which the reader is unfamiliar, a good reader will try to relate any of the author's details to their own familiar ground. A reader should use context clues to understand unfamiliar terminology, and recognize familiar information they have encountered in prior experience. Bringing prior life experience and knowledge to the test-taking situation is helpful in making connections. The ability to relate an unfamiliar idea to something the reader already knows is critical in understanding unique and new ideas.

When trying to make connections while reading, keep the following questions in mind:

- How does this feel familiar in personal experience?
- How is this similar to or different from other reading?
- How is this familiar in the real world?
- How does this relate to the world in general?

A reader should ask themself these questions during the act of reading in order to actively make connections to past and present experiences. Utilizing the ability to make connections is an important step in achieving synthesis.

Determining Importance in Reading

Being able to determine what is most important while reading is critical to synthesis. It is the difference between being able to tell what is necessary to full comprehension and that which is interesting but not necessary.

When determining the importance of an author's ideas, consider the following:

- Ask how critical an author's particular idea, assertion, or concept is to the overall message.

- Ask "is this an interesting fact or is this information essential to understanding the author's main idea?"

- Make a simple chart. On one side, list all of the important, essential points an author makes and on the other, list all of the interesting yet non-critical ideas.

- Highlight, circle, or underline any dates or data in non-fiction passages. Pay attention to headings, captions, and any graphs or diagrams.

- When reading a fictional passage, delineate important information such as theme, character, setting, conflict (what the problem is), and resolution (how the problem is fixed). Most often, these are the most important aspects contained in fictional text.

- If a non-fiction passage is instructional in nature, take physical note of any steps in the order of their importance as presented by the author. Look for words such as *first*, *next*, *then*, and *last*.

Determining the importance of an author's ideas is critical to synthesis in that it requires the test taker to parse out any unnecessary information and demonstrate they have the ability to make sound determination on what is important to the author, and what is merely a supporting or less critical detail.

Asking Questions While Reading

A reader must ask questions while reading. This demonstrates their ability to critically approach information and apply higher thinking skills to an author's content. Some of these questions have been addressed earlier in this section. A reader must ask what is or isn't important, what relates to their experience, and what relates to the world in general. However, it's important to ask other questions as well in order to make connections and synthesize reading material. Consider the following partial list of possibilities:

- What type of passage is this? Is it fiction? Non-fiction? Does it include data?

- Based on the type of passage, what information should be noted in order to make connections, visualize details, and determine importance?

- What is the author's message or theme? What is it they want the reader to understand?

- Is this passage trying to convince readers of something? What is it? If so, is the argument logical, convincing, and effective? How so? If not, how not?

- What do readers already know about this topic? Are there other viewpoints that support or contradict it?

- Is the information in this passage current and up to date?

- Is the author trying to teach readers a lesson? If so, what is it? Is there a moral to this story?

- How does this passage relate to experience?

- What is not as understandable in this passage? What context clues can help with understanding?

- What conclusions can be drawn? What predictions can be made?

Again, the above should be considered only a small example of the possibilities. Any question the reader asks while reading will help achieve synthesis and full reading comprehension.

Connections Between Different Texts

When analyzing two or more texts, there are several different aspects that need to be considered, particularly the styles (or the artful way in which the authors use diction to deliver a theme), points of view, and types of argument. In order to do so, one should compare and contrast the following elements between the texts:

- **Style**: narrative, persuasive, descriptive, informative, etc.
- **Tone**: sarcastic, angry, somber, humorous, etc.
- **Sentence structure**: simple (1 clause) compound (2 clauses), complex-compound (3 clauses)
- **Punctuation choice**: question marks, exclamation points, periods, dashes, etc.
- **Point of view**: first person, second person, third person

- **Paragraph structure**: long, short, both, differences between the two
- **Organizational structure**: compare/contrast, problem/solution, chronological, etc.

The following two poems and the essay concern the theme of death and are presented to demonstrate how to evaluate the above elements:

Poem 1:

How wonderful is Death,

Death, and his brother Sleep!

One, pale as yonder waning moon

With lips of lurid blue;

The other, rosy as the morn

When throned on ocean's wave

It blushes o'er the world;

Yet both so passing wonderful!

"Queen Mab," Percy Bysshe Shelley

Poem 2:

After great pain, a formal feeling comes –

The Nerves sit ceremonious, like Tombs –

The stiff Heart questions 'was it He, that bore,'

And 'Yesterday, or Centuries before'?

The Feet, mechanical, go round –

A Wooden way

Of Ground, or Air, or Ought –

Regardless grown,

A Quartz contentment, like a stone –

This is the Hour of Lead –

Remembered, if outlived,

As Freezing persons, recollect the Snow –

First – Chill – then Stupor – then the letting go –

"After Great Pain, A Formal Feeling Comes," Emily Dickinson

Essay 1

The Process of Dying

Death occurs in several stages. The first stage is the pre-active stage, which occurs a few days to weeks before death, in which the desire to eat and drink decreases, and the person may feel restless, irritable, and anxious. The second stage is the active stage, where the skin begins to cool, breathing becomes difficult as the lungs become congested (known as the "death rattle"), and the person loses control of their bodily fluids.

Once death occurs, there are also two stages. The first is clinical death, when the heart stops pumping blood and breathing ceases. This stage lasts approximately 4-6 minutes, and during this time, it is possible for a victim to be resuscitated via CPR or a defibrillator. After 6 minutes however, the oxygen stores within the brain begin to deplete, and the victim enters biological death. This is the point of no return, as the cells of the brain and vital organs begin to die, a process that is irreversible.

Now, using the outline above, the similarities and differences between the two passages are considered:

1. **Style**: The two poems are both descriptive as they focus on descriptions and sensations to convey their messages and do not follow any sort of timeline. The third selection is an expository style, presenting purely factual evidence on death, completely devoid of emotion.

2. **Tone**: Readers should notice the differences in the word choices between the two poems. Percy Shelley's word choices—"wonderful," "rosy," "blushes," "ocean"—surrounding death indicates that he views death in a welcoming manner as his words carry positive charges. The word choices by Dickinson, however, carry negative connotations—"pain," "wooden," "stone," "lead," "chill," "tombs"—which indicates an aversion to death. In contrast, the expository passage has no emotionally-charged words of any kind and seems to view death simply as a process that happens, neither welcoming nor fearing it. The tone in this passage, therefore, is neutral.

3. **Sentence structure**: Shelley's poem is composed mostly of compound sentences, which flow easily into one another. If read aloud, it sounds almost fluid, like the waves of the ocean he describes in his poem. His sentence structure mirrors the ease in which he views death. Dickinson's poem, on the other hand, is mostly simple sentences that are short and curt. They do not flow easily into one another, possibly representing her hesitancy and discomfort in her views of death. The expository passage contains many complex-compound sentences, which are used to accommodate lots of information. The structure of these sentences contributes to the overall informative nature of the selection.

4. **Punctuation choice**: Shelley uses commas, semicolons, and exclamation points in his poem, which, combined with his word choices and sentence structure, contributes to the overall positive tone of the poem. Dickinson uses lots of dashes, which make the poem feel almost cutting and jagged, which contributes to the overall negative tone of her poem. The expository text uses only commas and periods, which adds to the overall neutral tone of the selection.

5. **Point of view**: The point of view in all three selections is third person. In the two poems, there are no obvious pronouns; however, they both are presented in the third-person point of view, as Shelley speaks of Death in the third person, and Dickinson refers to "freezing persons." Generally, if there are no first- or second-person pronouns in a selection, the view is third person. The informational selection also uses third-person point of view, as it avoids any first- or second-person pronouns.

6. **Paragraph/stanza structure**: Shelley's poem is one stanza long, making it inherently simple in nature. The simplicity of the single stanza is representative (again) of the comfort in which the author finds the topic of death. Dickinson's poem is much lengthier, and comparatively, could signify the difficulty of letting go of the death of a loved one. The paragraph structure of the essay is much longer than the two and is used to fit in a lot more information than the poems, as the poems are trying to convey emotion, and the essay is presenting facts.

7. **Organizational structure**: Shelley's poem uses a compare and contrast method to illustrate the similarities between death and sleep: that death is merely a paler, bluer brother to the warm and rosy sleep. Dickinson's structure, however, is descriptive, focusing primarily on feelings and sensations. The expository passage, on the other hand, is chronologically-organized, as it follows a timeline of events that occur in stages.

When analyzing the different structures, it may be helpful to make a table and use single words to compare and contrast the texts:

Elements	Queen Mab	After Great Pain	Process of Dying
Style	Descriptive	Descriptive	Expository
Tone	Warm	Cold	Neutral
Sentence Structure	Fluid	Jagged	Long
Punctuation Choice	!	—	.
Point of View	Third	Third	Third
Paragraph Structure	Short	Longer	Longest
Organizational Structure	Compare-Contrast	Descriptive	Chronological

Using this table, the differences become very clear. Although the two poems are both about death, their word tone, sentence structure, punctuation choices, and organization depict differences in how the authors perceive death, while the elements in the expository text clearly indicate an objective view of death. It should be noted that these are only a handful of the endless possible interpretations the reader could make.

Analysis of History/Social Studies Excerpts

The ACCUPLACER will test for the ability to read substantial, historically based excerpts, and then answer comprehension questions based on content. The test taker will likely encounter U.S. history, or social science, passages within the test. One is likely to be from a U.S. founding document or work that has had great impact on history. The test may also include one or more passages from social sciences such as economics, psychology, or sociology.

For these types of questions, the test taker will need to utilize all the reading comprehension skills discussed above, but mastery of further skills will help. This section addresses those skills.

Comprehending Test Questions Prior to Reading

While preparing for a historical passage on a standardized test, first read the test questions, and then quickly scan the test answers prior to reading the passage itself. Notice there is a difference between the terms *read* and *scans*. Reading involves full concentration while addressing every word. Scanning involves quickly glancing at text in chunks, noting important dates, words, and ideas along the way. Reading test questions will help the test taker know what information to focus on in the historical passage. Scanning answers will help the test taker focus on possible answer options while reading the passage.

When reading standardized test questions that address historical passages, be sure to clearly understand what each question is asking. Is a question asking about vocabulary? Is another asking for the test taker to find a specific historical fact? Do any of the questions require the test taker to draw conclusions, identify an author's topic, tone, or position? Knowing what content to address will help the test taker focus on the information they will be asked about later. However, the test taker should approach this reading comprehension technique with some caution. It is tempting to only look for the right answers within any given passage. Do not put on "reading blinders" and ignore all other information presented in a passage. It is important to fully read every passage and not just scan it. Strictly looking for what may be the right answers to test questions can cause the test taker to ignore important contextual clues that actually require critical thinking in order to identify correct answers. Scanning a passage for what appears to be wrong answers can have a similar result.

When reading test questions prior to tackling a historical passage, be sure to understand what skills the test is assessing, and then fully read the related passage with those skills in mind. Focus on every word in both the test questions and the passage itself. Read with a critical eye and a logical mind.

Reading for Factual Information

Standardized test questions that ask for factual information are usually straightforward. These types of questions will either ask the test taker to confirm a fact by choosing a correct answer, or to select a correct answer based on a negative fact question.

For example, the test taker may encounter a passage from Lincoln's Gettysburg address. A corresponding test question may ask the following:

> Which war is Abraham Lincoln referring to in the following passage?: "Now we are engaged in a great civil war, testing whether that nation, or any nation so conceived and so dedicated, can long endure."

This type of question is asking the test taker to confirm a simple fact. Given options such as World War I, the War of Spanish Succession, World War II, and the American Civil War, the test taker should be able to correctly identify the American Civil War based on the words "civil war" within the passage itself, and, hopefully, through general knowledge. In this case, reading the test question and scanning answer options ahead of reading the Gettysburg address would help quickly identify the correct answer. Similarly, a test taker may be asked to confirm a historical fact based on a negative fact question. For example, a passage's corresponding test question may ask the following:

> Which option is incorrect based on the above passage?

Given a variety of choices speaking about which war Abraham Lincoln was addressing, the test taker would need to eliminate all correct answers pertaining to the American Civil War and choose the answer

choice referencing a different war. In other words, the correct answer is the one that contradicts the information in the passage.

It is important to remember that reading for factual information is straightforward. The test taker must distinguish fact from bias. Factual statements can be proven or disproven independent of the author and from a variety of other sources. Remember, successfully answering questions regarding factual information may require the test taker to re-read the passage, as these types of questions test for attention to detail.

Analysis of Science Excerpts

The ACCUPLACER usually contains at includes at least two science passages that address the fundamental concepts of Earth science, biology, chemistry, and/or physics. While prior general knowledge of these subjects is helpful in determining correct test answers, the test taker's ability to comprehend the passages is key to success. When reading scientific excerpts, the test taker must be able to examine quantitative information, identify hypotheses, interpret data, and consider implications of the material they are presented with. It is helpful, at this point, to reference the above section on comprehending test questions prior to reading. The same rules apply: read questions and scan questions, along with their answers, prior to fully reading a passage. Be informed prior to approaching a scientific text. A test taker should know what they will be asked and how to apply their reading skills. In this section of the test, it is also likely that a test taker will encounter graphs and charts to assess their ability to interpret scientific data with an appropriate conclusion. This section will determine the skills necessary to address scientific data presented through identifying hypotheses, through reading and examining data, and through interpreting data representation passages.

Examine Hypotheses

When presented with fundamental, scientific concepts, it is important to read for understanding. The most basic skill in achieving this literacy is to understand the concept of hypothesis and moreover, to be able to identify it in a particular passage. A hypothesis is a proposed idea that needs further investigation in order to be proven true or false. While it can be considered an educated guess, a hypothesis goes more in depth in its attempt to explain something that is not currently accepted within scientific theory. It requires further experimentation and data gathering to test its validity and is subject to change, based on scientifically conducted test results. Being able to read a science passage and understand its main purpose, including any hypotheses, helps the test taker understand data-driven evidence. It helps the test taker to be able to correctly answer questions about the science excerpt they are asked to read.

When reading to identify a hypothesis, a test taker should ask, "What is the passage trying to establish? What is the passage's main idea? What evidence does the passage contain that either supports or refutes this idea?" Asking oneself these questions will help identify a hypothesis. Additionally, hypotheses are logical statements that are testable and use very precise language.

Review the following hypothesis example:

> Consuming excess sugar in the form of beverages has a greater impact on childhood obesity and subsequent weight gain than excessive sugar from food.

While this is likely a true statement, it is still only a conceptual idea in a text passage regarding how sugar consumption affects childhood obesity, unless the passage also contains tested data that either

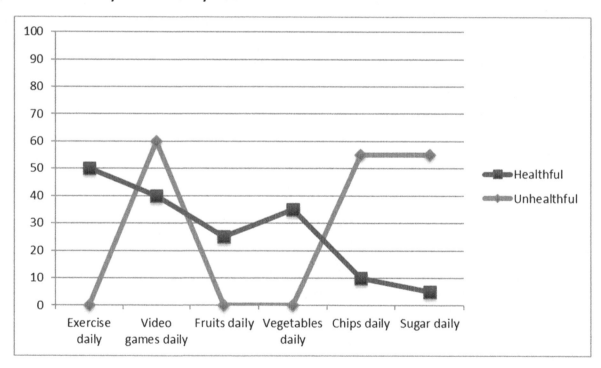

proves or disproves the statement. A test taker could expect the rest of the passage to cite data proving that children who drink empty calories and don't exercise will, in fact, be obese.

A hypothesis goes further in that, given its ability to be proven or disproven, it may result in further hypotheses that require extended research. For example, the hypothesis regarding sugar consumption in drinks, after undergoing rigorous testing, may lead scientists to state another hypothesis such as the following:

> Consuming excess sugar in the form of beverages as opposed to food items is a habit found in mostly sedentary children.

This new, working hypothesis further focuses not just on the source of an excess of calories, but tries an "educated guess" that empty caloric intake has a direct, subsequent impact on physical behavior.

The data-driven chart below is similar to an illustration a test taker might see in relation to the hypothesis on sugar consumption in children:

Behaviors of Healthy and Unhealthy Kids

While this guide will address other data-driven passages a test taker could expect to see within a given science excerpt, note that the hypothesis regarding childhood sugar intake and rate of exercise has undergone scientific examination and yielded results that support its truth.

When reading a science passage to determine its hypothesis, a test taker should look for a concept that attempts to explain a phenomenon, is testable, logical, precisely worded, and yields data-driven results. The test taker should scan the presented passage for any word or data-driven clues that will help identify the hypothesis, and then be able to correctly answer test questions regarding the hypothesis based on their critical thinking skills.

Interpreting Data and Considering Implications

The ACCUPLACER is likely to contain one or more data-driven science passages that require the test taker to examine evidence within a particular type of graphic. The test taker will then be required to interpret the data and answer questions demonstrating their ability to draw logical conclusions.

In general, there are two types of data: qualitative and quantitative. Science passages may contain both, but simply put, quantitative data is reflected numerically and qualitative is not. Qualitative data is based on its qualities. In other words, qualitative data tends to present information more in subjective generalities (for example, relating to size or appearance). Quantitative data is based on numerical findings such as percentages. Quantitative data will be described in numerical terms. While both types of data are valid, the test taker will more likely be faced with having to interpret quantitative data through one or more graphic(s), and then be required to answer questions regarding the numerical data. The section of this study guide briefly addresses how data may be displayed in line graphs, bar charts, circle graphs, and scatter plots. A test taker should take the time to learn the skills it takes to interpret quantitative data. An example of a line graph is as follows:

Cell Phone Use in Kiteville, 2000-2006

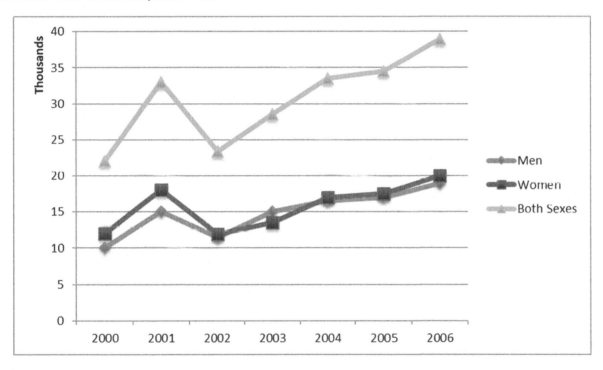

A line graph presents quantitative data on both horizontal (side to side) and vertical (up and down) axes. It requires the test taker to examine information across varying data points. When reading a line graph, a test taker should pay attention to any headings, as these indicate a title for the data it contains. In the above example, the test taker can anticipate the line graph contains numerical data regarding the use of cellphones during a certain time period. From there, a test taker should carefully read any outlying words or phrases that will help determine the meaning of data within the horizontal and vertical axes. In this example, the vertical axis displays the total number of people in increments of 5,000. Horizontally, the graph displays yearly markers, and the reader can assume the data presented accounts for a full calendar year. In addition, the line graph also defines its data points by shapes. Some data points represent the number of men. Some data points represent the number of women, and a third type of data point represents the number of both sexes combined.

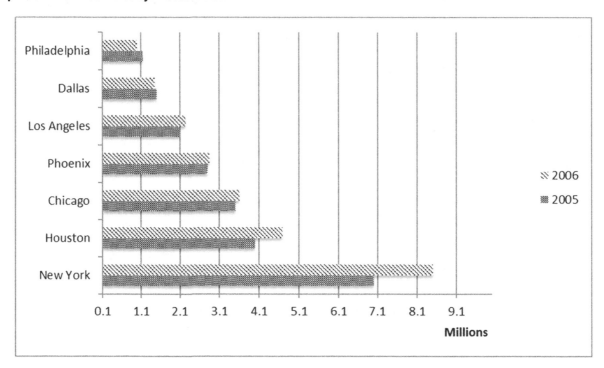

A test taker may be asked to read and interpret the graph's data, then answer questions about it. For example, the test may ask, *In which year did men seem to decrease cellphone use?* then require the test taker to select the correct answer. Similarly, the test taker may encounter a question such as *Which year yielded the highest number of cellphone users overall?* The test taker should be able to identify the correct answer as 2006.

A **bar graph** presents quantitative data through the use of lines or rectangles. The height and length of these lines or rectangles corresponds to the magnitude of the numerical data for that particular category or attribute. The data presented may represent information over time, showing shaded data over time or over other defined parameters. A bar graph will also utilize horizontal and vertical axes. An example of a bar graph is as follows:

Population Growth in Major U.S. Cities

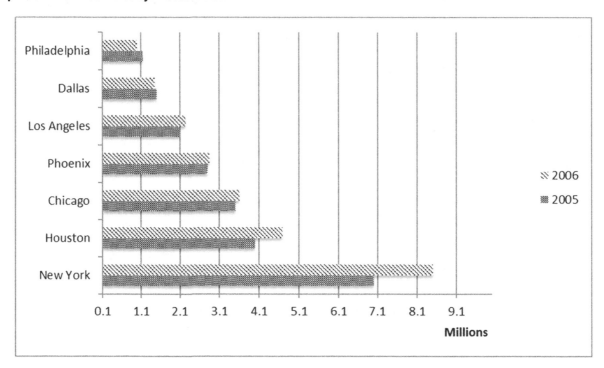

Reading the data in a bar graph is similar to the skills needed to read a line graph. The test taker should read and comprehend all heading information, as well as information provided along the horizontal and vertical axes. Note that the graph pertains to the population of some major U.S. cities. The "values" of these cities can be found along the left side of the graph, along the vertical axis. The population values can be found along the horizontal axes. Notice how the graph uses shaded bars to depict the change in population over time, as the heading indicates. Therefore, when the test taker is asked a question such as, *Which major U.S. city experienced the greatest amount of population growth during the depicted two year cycle,* the reader should be able to determine a correct answer of New York. It is important to pay particular attention to color, length, data points, and both axes, as well as any outlying header information in order to be able to answer graph-like test questions.

A circle graph presents quantitative data in the form of a circle (also sometimes referred to as a pie chart). The same principles apply: the test taker should look for numerical data within the confines of the circle itself but also note any outlying information that may be included in a header, footer, or to the side of the circle. A circle graph will not depict horizontal or vertical axis information, but will instead rely on the reader's ability to visually take note of segmented circle pieces and apply information accordingly. An example of a circle graph is as follows:

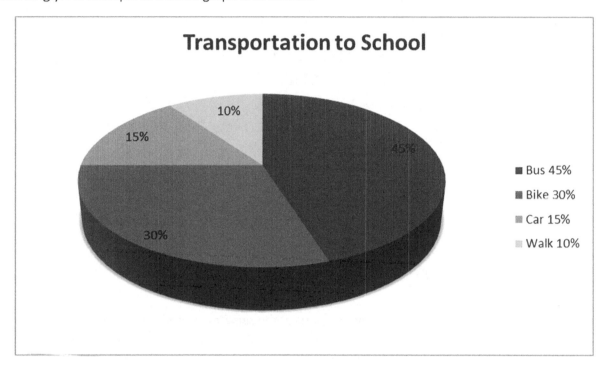

Notice the heading "Transportation to School." This should indicate to the test taker that the topic of the circle graph is how people traditionally get to school. To the right of the graph, the reader should comprehend that the data percentages contained within it directly correspond to the method of transportation. In this graph, the data is represented through the use shades and pattern. Each transportation method has its own shade. For example, if the test taker was then asked, *Which method of school transportation is most widely utilized,* the reader should be able to identify school bus as the correct answer.

Be wary of test questions that ask test takers to draw conclusions based on information that is not present. For example, it is not possible to determine, given the parameters of this circle graph, whether the population presented is of a particular gender or ethnic group. This graph does not represent data from a particular city or school district. It does not distinguish between student grade levels and, although the reader could infer that the typical student must be of driving age if cars are included, this is not necessarily the case. Elementary school students may rely on parents or others to drive them by personal methods. Therefore, do not read too much into data that is not presented. Only rely on the quantitative data that is presented in order to answer questions.

A scatter plot or scatter diagram is a graph that depicts quantitative data across plotted points. It will involve at least two sets of data. It will also involve horizontal and vertical axes.

An example of a scatter plot is as follows:

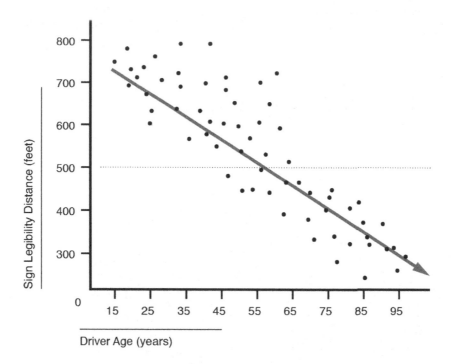

The skills needed to address a scatter plot are essentially the same as in other graph examples. Note any topic headings, as well as horizontal or vertical axis information. In the sample above, the reader can determine the data addresses a driver's ability to correctly and legibly read road signs as related to their age. Again, note the information that is absent. The test taker is not given the data to assess a time period, location, or driver gender. It simply requires the reader to note an approximate age to the ability to correctly identify road signs from a distance measured in feet. Notice that the overall graph also displays a trend. In this case, the data indicates a negative one and possibly supports the hypothesis that as a driver ages, their ability to correctly read a road sign at over 500 feet tends to decline over time. If the test taker were to be asked, *At what approximation in feet does a sixteen-year-old driver correctly see and read a street sign,* the answer would be the option closest to 700 feet.

Reading and examining scientific data in excerpts involves all of a reader's contextual reading, data interpretation, drawing logical conclusions based only on the information presented, and their application of critical thinking skills across a set of interpretive questions. Thorough comprehension and attention to detail is necessary to achieve test success.

Vocabulary

Identifying Roots

By analyzing and understanding Latin, Greek, and Anglo-Saxon word roots and structure, authors better convey the thoughts they want to express to the readers of their words and help them to determine their meanings within the flow and without their missing a beat. For instance, **context**—how words are

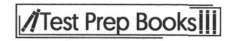

used in sentences—is from the Latin for *contextus*, which means "together" + "to weave," and gives readers a graphic for the minds' eyes to see the coming together of their usage. Like every other topic discussed herein, context is needed for understanding. This element actually has a second, crucial meaning. Context is not only the *how*, but the revealed moment of the *why* a writing has been composed; it is the "Aha" moment.

The way *how* words are used in sentences is important because it also gives meaning and cohesion from sentence to sentence, paragraph to paragraph, and page after page. In other words, it gives the document continuity.

Another upside of the how side is that readers have opportunities to understand new words with which they are unfamiliar. Of course, people can always look words up if a dictionary or thesaurus, if available, but meaning might be gleaned on the spot in a piece that is well-written. **Synonyms** (words or phrases that mean about the same) and **antonyms** (words or phrases that mean the opposite of the specific word) in context give clues to meanings, and sometimes reiteration of a word might add clarification. Repetition, wisely used, can also serve as a part of how a piece flows.

The revealed moment of the *why* is important because context, up to that moment, has determined the shape of the text. This is, essentially, to bring out what it is all about.

Prefixes

A **prefix** is a word, letter, or number that is placed before another. It adjusts or qualifies the original word's meaning.

Four prefixes represent 97 percent of English words with prefixes. They are:

- *dis-* means "not" or "opposite of"; *dis*abled
- *in-, im-, il-, ir-* mean "not"; *il*literate
- *re-* means "again"; *re*turn
- *un-* means "not"; *un*predictable

Other commons prefixes include:

- *anti-* means "against"; antibacterial
- *fore-* means "before"; forefront
- *mis-* means "wrongly"; misunderstand
- *non-* means "not"; nonsense
- *over-* means "over"; overabundance
- *pre-* means "before"; preheat
- *super-* means "above"; superman

Suffixes

The official definition of a **suffix** is "a morpheme added at the end of a word to form a derivative." In English, that means a suffix is a letter or group of letters added at the end of a word to form another word. The word created with the addition is either a different tense of the same word (*help + ed = helped*) or a new word (*help + ful = helpful*).

They are:

- *-ed* is used to make present tense verbs into past tense verbs; wash*ed*
- *-ing* is used to make a present tense verb into a present participle verb; wash*ing*
- *-ly* is used to make characteristic of; love*ly*
- *-s* or *–es* are used to make more than one; chair*s* or box*es*

Other common suffixes include:

- *-able* means can be done; deplor*eable*
- *-al* means having characteristics of; comic*al*
- *-est* means comparative; great*est*
- *-ful* means full of; wonder*ful*
- *-ism* means belief in; commun*eism*
- *-less* means without; faith*less*
- *-ment* means action or process; accomplish*ment*
- *-ness* means state of; happ*yiness*
- *-ize* means to render, to make; terror*ize,* steril*ize*
- *-ise* means ditto, only this is primarily the British variant of *–ize;* surpr*ise,* advert*ise*
- -ced means go; spelling variations include -cede (concede, recede); -ceed (only three: proceed, exceed, succeed); -sede (the only one: supersede)

(Note: In some of the examples above, the *e* has been deleted.)

Finding the Meaning of Words and Phrases in Context

There will be many occasions in one's reading career in which an unknown word or a word with multiple meanings will pop up. There are ways of determining what these words or phrases mean that do not require the use of the dictionary, which is especially helpful during a test where one may not be available. Even outside of the exam, knowing how to derive an understanding of a word via **context clues** will be a critical skill in the real world. The context is the circumstances in which a story or a passage is happening, and can usually be found in the series of words directly before or directly after the word or phrase in question. The clues are the words that hint towards the meaning of the unknown word or phrase. The author may use synonyms or antonyms that you can use. **Synonyms** refer to words that have the same meaning as another word (e.g., instructor/teacher/educator, canine/dog, feline/cat, herbivore/vegetarian). **Antonyms** refer to words that have the opposite meaning as another word (e.g., true/false, up/down, in/out, right/wrong).

There may be questions that ask about the meaning of a particular word or phrase within a passage. There are a couple ways to approach these kinds of questions:

- Define the word or phrase in a way that is easy to comprehend (using context clues).
- Try out each answer choice in place of the word.

To demonstrate, here's an example from *Alice in Wonderland*:

Alice was beginning to get very tired of sitting by her sister on the bank, and of having nothing to do: once or twice she <u>peeped</u> into the book her sister was reading, but it had no pictures or conversations in it, "and what is the use of a book," thought Alice, "without pictures or conversations?"

Q: As it is used in the selection, the word <u>peeped</u> means:

Using the first technique, before looking at the answers, define the word "peeped" using context clues and then find the matching answer. Then, analyze the entire passage in order to determine the meaning, not just the surrounding words.

To begin, imagine a blank where the word should be and put a synonym or definition there: "once or twice she ___ into the book her sister was reading." The context clue here is the book. It may be tempting to put "read" where the blank is, but notice the preposition word, "into." One does not read *into* a book, one simply reads a book, and since reading a book requires that it is seen with a pair of eyes, then "look" would make the most sense to put into the blank: "once or twice she <u>looked</u> into the book her sister was reading."

Once an easy-to-understand word or synonym has been supplanted, check to make sure it makes sense with the rest of the passage. What happened after she looked into the book? She thought to herself how a book without pictures or conversations is useless. This situation in its entirety makes sense.

Now check the answer choices for a match:
   a. To make a high-pitched cry
   b. To smack
   c. To look curiously
   d. To pout

Since the word was already defined, answer choice (c) is the best option.

Using the second technique, replace the figurative blank with each of the answer choices and determine which one is the most appropriate. Remember to look further into the passage to clarify that they work, because they could still make sense out of context.

Once or twice she <u>made a high pitched cry</u> into the book her sister was reading.

Once or twice she <u>smacked</u> the book her sister was reading.

Once or twice she <u>looked curiously</u> into the book her sister was reading.

Once or twice she <u>pouted</u> into the book her sister was reading.

For Choice *A*, it does not make much sense in any context for a person to yell into a book, unless maybe something terrible has happened in the story. Given that afterward Alice thinks to herself how useless a book without pictures is, this option does not make sense within context.

For Choice *B*, smacking a book someone is reading may make sense if the rest of the passage indicates there a reason for doing so. If Alice was angry or her sister had shoved it in her face, then maybe smacking the book would make sense within context. However, since whatever she does with the book causes her to think, "what is the use of a book without pictures or conversations?" then answer Choice *B* is not an appropriate answer.

Answer Choice *C* fits well within context, given her subsequent thoughts on the matter.

Answer Choice *D* does not make sense in context or grammatically, as people do not "pout into" things.

This is a simple example to illustrate the techniques outlined above. There may, however, be a question in which all of the definitions are correct and also make sense out of context, in which the appropriate context clues will really need to be honed in on in order to determine the correct answer. For example, here is another passage from *Alice in Wonderland*:

> ... but when the Rabbit actually took a watch out of its waistcoat pocket, and looked at it, and then hurried on, Alice <u>started</u> to her feet, for it flashed across her mind that she had never before seen a rabbit with either a waistcoat-pocket or a watch to take out of it, and burning with curiosity, she ran across the field after it, and was just in time to see it pop down a large rabbit-hole under the hedge.

Q: As it is used in the passage, the word <u>started</u> means:
   a. To turn on
   b. To begin
   c. To move quickly
   d. To be surprised

All of these words qualify as a definition of start, but using context clues, the correct answer can be identified using one of the two techniques above. It's easy to see that one does not turn on, begin, or be surprised to one's feet. The selection also states that she "ran across the field after it," indicating that she was in a hurry. Therefore, to move quickly would make the most sense in this context.

The same strategies can be applied to vocabulary that may be completely unfamiliar. In this case, focus on the words before or after the unknown word in order to determine its definition. Take this sentence, for example:

"Sam was such a <u>miser</u> that he forced Andrew to pay him twelve cents for the candy, even though he had a large inheritance and he knew his friend was poor."

Unlike with assertion questions, for vocabulary questions, it may be necessary to apply some critical thinking skills that may not be explicitly stated within the passage. Think about the implications of the passage, or what the text is trying to say. With this example, it is important to realize that it is considered unusually stingy for a person to demand so little money from someone instead of just letting their friend have the candy, especially if this person is already wealthy. Hence, a <u>miser</u> is a greedy or stingy individual.

Questions about complex vocabulary may not be explicitly asked, but this is a useful skill to know. If there is an unfamiliar word while reading a passage and its definition goes unknown, it is possible to miss out on a critical message that could inhibit the ability to appropriately answer the questions. Practicing this technique in daily life will sharpen this ability to derive meanings from context clues with ease.

Practice Questions

Questions 1-5 are based upon the following passage:

The Myth of Head Heat Loss

It has recently been brought to my attention that most people believe that 75% of your body heat is lost through your head. I had certainly heard this before, and am not going to attempt to say I didn't believe it when I first heard it. It is natural to be gullible to anything said with enough authority. But the "fact" that the majority of your body heat is lost through your head is a lie.

Let me explain. Heat loss is proportional to surface area exposed. An elephant loses a great deal more heat than an anteater because it has a much greater surface area than an anteater. Each cell has mitochondria that produce energy in the form of heat, and it takes a lot more energy to run an elephant than an anteater.

So, each part of your body loses its proportional amount of heat in accordance with its surface area. The human torso probably loses the most heat, though the legs lose a significant amount as well. Some people have asked, "Why does it feel so much warmer when you cover your head than when you don't?" Well, that's because your head, because it is not clothed, is losing a lot of heat while the clothing on the rest of your body provides insulation. If you went outside with a hat and pants but no shirt, not only would you look stupid but your heat loss would be significantly greater because so much more of you would be exposed. So, if given the choice to cover your chest or your head in the cold, choose the chest. It could save your life.

1. What is the primary purpose of this passage?
    a. To provide evidence that disproves a myth
    b. To compare elephants and anteaters
    c. To explain why it is appropriate to wear clothes in winter
    d. To show how people are gullible

2. Which of the following best describes the main idea of the passage?
    a. It is better to wear a shirt than a hat
    b. Heat loss is proportional to surface area exposed
    c. It is natural to be gullible
    d. The human chest loses the most heat

3. Why does the author compare elephants and anteaters?
    a. To express an opinion
    b. To give an example that helps clarify the main point
    c. To show the differences between them
    d. To persuade why one is better than the other

4. Which of the following best describes the tone of the passage?
    a. Harsh
    b. Angry
    c. Casual
    d. Indifferent

5. Which of the following sentences provides the best evidence to support the main idea?
   a. "It is natural to be gullible to anything said with enough authority."
   b. "Each part of your body loses its proportional amount of heat in accordance with its surface area."
   c. "If given the choice to cover your chest or your head in the cold, choose the chest."
   d. "But the 'fact' that the majority of your body heat it lost through your head is a lie."

Answer Explanations

1. A: Not only does the article provide examples to disprove a myth, the title also suggests that the article is trying to disprove a myth. Further, the sentence, "But the 'fact' that the majority of your body heat is lost through your head is a lie," and then the subsequent "let me explain," demonstrates the author's intention in disproving a myth. *B* is incorrect because although the selection does compare elephants and anteaters, it does so in order to prove a point, and is not the primary reason that the selection was written. *C* is incorrect because even though the article mentions somebody wearing clothes in the winter, and that doing so could save your life, wearing clothes in the winter is not the primary reason this article was written. *D* is incorrect because the article only mentions that people are gullible once, and makes no further comment on the matter, so this cannot be the primary purpose.

2. B: If the myth is that most of one's body heat is lost through their head, then the fact that heat loss is proportional to surface area exposed is the best evidence that disproves it, since one's head is a great deal less surface area than the rest of the body, making *B* the correct choice. "It is better to wear a shirt than a hat" does not provide evidence that disproves the fact that the head loses more heat than the rest of the body. Thus, *A* is incorrect. *C* is incorrect because gullibility is mentioned only once in this passage and the rest of the article ignores this statement, so clearly it is not the main idea. Finally, *D* is incorrect because though the article mentions that the human chest probably loses the most heat, it is to provide an example of the evidence that heat loss is proportional to surface area exposed, so this is not the main idea of the passage.

3. B: Choice *B* is correct because the author is trying to demonstrate the main idea, which is that heat loss is proportional to surface area, and so they compare two animals with different surface areas to clarify the main point. Choice *A* is incorrect because the author uses elephants and anteaters to prove a point that heat loss is proportional to surface area, not to express an opinion. Choice *C* is incorrect because though the author does use them to show differences, they do so in order to give examples that prove the above points. Choice *D* is incorrect because there is no language to indicate favoritism between the two animals.

4. C: Because of the way the author addresses the reader and the colloquial language the author uses (i.e., "let me explain," "so," "well," didn't," "you would look stupid," etc.), Choice *C* is the best answer because it has a much more casual tone than the usual informative article. Choice *A* may be a tempting choice because the author says the "fact" that most of one's heat is lost through their head is a "lie" and that someone who does not wear a shirt in the cold looks stupid. However, this only happens twice within the passage, and the passage does not give an overall tone of harshness. Choice *B* is incorrect because again, while not necessarily nice, the language does not carry an angry charge. The author is clearly not indifferent to the subject because of the passionate language that they use, so Choice *D* is incorrect.

5. B: Choice *B* is correct. The primary purpose of the article is to provide evidence to disprove the myth that most of a person's heat is lost through their head. The fact that each part of the body loses heat in proportion to its surface area is the best evidence to disprove this myth. Choice *A* is incorrect because again, gullibility is not a main contributor to this article, but it may be common to see questions on the test that give the same wrong answer in order to try and trick the test taker. Choice *C* only suggests what you should do with this information; it is not the primary evidence itself. Choice *D*, while tempting, is actually not evidence. It does not give any reason for why it is a lie; it simply states that it is. Evidence is factual information that supports a claim.

Expression of Ideas

Development

for words that are close together with the same (or similar) meanings.

Proposition

The **proposition** (also called the **claim** since it can be true or false) is a clear statement of the point or idea the writer is trying to make. The length or format of a proposition can vary, but it often takes the form of a **topic sentence**. A good topic sentence is:

- Clear: does not weave a complicated web of words for the reader to decode or unwrap

- Concise: presents only the information needed to make the claim and doesn't clutter up the statement with unnecessary details

- Precise: clarifies the exact point the writer wants to make and doesn't use broad, overreaching statements

Look at the following example:

> The civil rights movement, from its genesis in the Emancipation Proclamation to its current struggles with de facto discrimination, has changed the face of the United States more than any other factor in its history.

Is the statement clear? Yes, the statement is fairly clear, although other words can be substituted for "genesis" and "de facto" to make it easier to understand.

Is the statement concise? No, the statement is not concise. Details about the Emancipation Proclamation and the current state of the movement are unnecessary for a topic sentence. Those details should be saved for the body of the text.

Is the statement precise? No, the statement is not precise. What exactly does the writer mean by "changed the face of the United States"? The writer should be more specific about the effects of the movement. Also, suggesting that something has a greater impact than anything else in U.S. history is far too ambitious a statement to make.

A better version might look like this:

> The civil rights movement has greatly increased the career opportunities available for Black Americans.

The unnecessary language and details are removed, and the claim can now be measured and supported.

Support

Once the main idea or proposition is stated, the writer attempts to prove or **support** the claim with text evidence and supporting details.

Take for example the sentence, "Seat belts save lives." Though most people can't argue with this statement, its impact on the reader is much greater when supported by additional content. The writer can support this idea by:

- Providing statistics on the rate of highway fatalities alongside statistics for estimated seat belt usage.

- Explaining the science behind a car accident and what happens to a passenger who doesn't use a seat belt.

- Offering anecdotal evidence or true stories from reliable sources on how seat belts prevent fatal injuries in car crashes.

However, using only one form of supporting evidence is not nearly as effective as using a variety to support a claim. Presenting only a list of statistics can be boring to the reader, but providing a true story that's both interesting and humanizing helps. In addition, one example isn't always enough to prove the writer's larger point, so combining it with other examples is extremely effective for the writing. Thus, when reading a passage, don't just look for a single form of supporting evidence.

Another key aspect of supporting evidence is a *reliable source*. Does the writer include the source of the information? If so, is the source well known and trustworthy? Is there a potential for bias? For example, a seat belt study done by a seat belt manufacturer may have its own agenda to promote.

Focus

Good writing stays focused and on topic. During the test, determine the main idea for each passage and then look for times when the writer strays from the point they're trying to make. Let's go back to the seat belt example. If the writer suddenly begins talking about how well airbags, crumple zones, or other safety features work to save lives, they might be losing focus from the topic of "safety belts."

Focus can also refer to individual sentences. Sometimes the writer does address the main topic, but in a confusing way. For example:

> Thanks to seat belt usage, survival in serious car accidents has shown a consistently steady increase since the development of the retractable seat belt in the 1950s.

This statement is definitely on topic, but it's not easy to follow. A simpler, more focused version of this sentence might look like this:

> Seat belts have consistently prevented car fatalities since the 1950s.

Providing adequate information is another aspect of focused writing. Statements like "seat belts are important" and "many people drive cars" are true, but they're so general that they don't contribute much to the writer's case. When reading a passage, watch for these kinds of unfocused statements.

Organization

Good writing is not merely a random collection of sentences. No matter how well written, sentences must relate and coordinate appropriately with one another. If not, the writing seems random, haphazard, and disorganized. Therefore, good writing must be organized, where each sentence fits a larger context and relates to the sentences around it.

Logical Sequence

Even if the writer includes plenty of information to support their point, the writing is only effective when the information is in a logical order. **Logical sequencing** is really just common sense, but it's an important writing technique. First, the writer should introduce the main idea, whether for a paragraph, a section, or the entire piece. Second, they should present evidence to support the main idea by using transitional language. This shows the reader how the information relates to the main idea and to the sentences around it. The writer should then take time to interpret the information, making sure necessary connections are obvious to the reader. Finally, the writer can summarize the information in a closing section.

Although most writing follows this pattern, it isn't a set rule. Sometimes writers change the order for effect. For example, the writer can begin with a surprising piece of supporting information to grab the reader's attention, and then transition to the main idea. Thus, if a passage doesn't follow the logical order, don't immediately assume it's wrong. However, most writing usually settles into a logical sequence after a nontraditional beginning.

Introductions and Conclusions

Examining the writer's strategies for introductions and conclusions puts the reader in the right mindset to interpret the rest of the passage. Look for methods the writer might use for introductions such as:

- Stating the main point immediately, followed by outlining how the rest of the piece supports this claim.

- Establishing important, smaller pieces of the main idea first, and then grouping these points into a case for the main idea.

- Opening with a quotation, anecdote, question, seeming paradox, or other piece of interesting information, and then using it to lead to the main point.

Whatever method the writer chooses, the introduction should make their intention clear, establish their voice as a credible one, and encourage a person to continue reading.

Conclusions tend to follow a similar pattern. In them, the writer restates their main idea a final time, often after summarizing the smaller pieces of that idea. If the introduction uses a quote or anecdote to grab the reader's attention, the conclusion often makes reference to it again. Whatever way the writer chooses to arrange the conclusion, the final restatement of the main idea should be clear and simple for the reader to interpret.

Finally, conclusions shouldn't introduce any new information.

Transition Words

The writer should act as a guide, showing the reader how all the sentences fit together. Consider this example concerning seat belts:

> Seat belts save more lives than any other automobile safety feature. Many studies show that airbags save lives as well. Not all cars have airbags. Many older cars don't. Air bags aren't entirely reliable. Studies show that in 15 percent of accidents, airbags don't deploy as designed. Seat belt malfunctions are extremely rare.

There's nothing wrong with any of these sentences individually, but together they're disjointed and difficult to follow. The best way for the writer to communicate information is through the use of transition words. Here are examples of transition words and phrases that tie sentences together, enabling a more natural flow:

- To show causality: as a result, therefore, and consequently
- To compare and contrast: *however, but,* and *on the other hand*
- To introduce examples: *for example, namely,* and *including*
- To show order of importance: *foremost, primarily, secondly,* and *lastly*

Note that this is not a complete list of transitions. There are many more that can be used; however, most fit into these or similar categories. The important point is that the words should clearly show the relationship between sentences, supporting information, and the main idea.

Here is an update to the previous example using transition words. These changes make it easier to read and bring clarity to the writer's points:

> Seat belts save more lives than any other automobile safety feature. Many studies show that airbags save lives as well; however, not all cars have airbags. For example, some older cars don't. Furthermore, air bags aren't entirely reliable. For example, studies show that in 15 percent of accidents, airbags don't deploy as designed, but, on the other hand, seat belt malfunctions are extremely rare.

Also, be prepared to analyze whether the writer is using the best transition word or phrase for the situation. Take this sentence for example: "As a result, seat belt malfunctions are extremely rare." This sentence doesn't make sense in the context above because the writer is trying to show the contrast between seat belts and airbags, not the causality.

Forming Paragraphs

A good **paragraph** should have the following characteristics:

- Be logical with organized sentences
- Have a *unified* purpose within itself
- Use sentences as *building blocks*
- Be a *distinct section* of a piece of writing
- Present a *single theme* introduced by a *topic sentence*
- Maintain a *consistent flow* through subsequent, relevant, well-placed sentences
- *Tell a story* of its own or have its own purpose, yet connect with what is written before and after
- Enlighten, entertain, and/or inform

Though certainly not set in stone, the length should be a consideration for the reader's sake, not merely for the sake of the topic. When paragraphs are especially short, the reader might experience an irregular, uneven effect; when they're much longer than 250 words, the reader's attention span, and probably their retention, is challenged. While a paragraph can technically be a sentence long, a good rule of thumb is for paragraphs to be at least three sentences long and no more than ten sentence long. An optimal word length is 100 to 250 words.

Relevance of Content

A reader must be able to evaluate the argument or point the author is trying to make and determine if it is adequately supported. The first step is to determine the main idea. The main idea is what the author wants to say about a specific topic. The next step is to locate the supporting details. An author uses supporting details to illustrate the main idea. These are the details that provide evidence or examples to help make a point. Supporting details often appear in the form of quotations, paraphrasing, or analysis. Test takers should then examine the text to make sure the author connects details and analysis to the main point. These steps are crucial to understanding the text and evaluating how well the author presents his or her argument and evidence. The following graphic demonstrates the connection between the main idea and the supporting details.

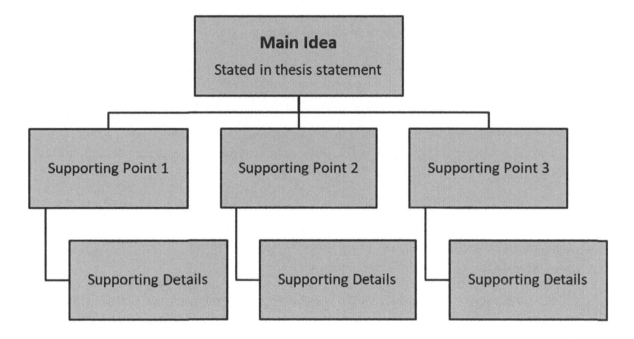

It is important to evaluate the author's supporting details to be sure that they are credible, provide evidence of the author's point, and directly support the main idea. Critical readers examine the facts used to support an author's argument and check those facts against other sources to be sure the facts are correct. They also check the validity of the sources used to be sure those sources are credible, academic, and/or peer- reviewed. A strong argument uses valid, measurable facts to support ideas.

Effective Language Use

Precision

People often think of **precision** in terms of math, but precise word choice is another key to successful writing. Since language itself is imprecise, it's important for the writer to find the exact word or words to convey the full, intended meaning of a given situation. For example:

> The number of deaths has gone down since seat belt laws started.

There are several problems with this sentence. First, the word *deaths* is too general. From the context, it's assumed that the writer is referring only to *deaths* caused by car accidents. However, without clarification, the sentence lacks impact and is probably untrue. The phrase "gone down" might be

accurate, but a more precise word could provide more information and greater accuracy. Did the numbers show a slow and steady decrease of highway fatalities or a sudden drop? If the latter is true, the writer is missing a chance to make their point more dramatically. Instead of "gone down" they could substitute *plummeted*, *fallen drastically*, or *rapidly diminished* to bring the information to life. Also, the phrase "seat belt laws" is unclear. Does it refer to laws requiring cars to include seat belts or to laws requiring drivers and passengers to use them? Finally, *started* is not a strong verb. Words like *enacted* or *adopted* are more direct and make the content more real. When put together, these changes create a far more powerful sentence:

> The number of highway fatalities has plummeted since laws requiring seat belt usage were enacted.

However, it's important to note that precise word choice can sometimes be taken too far. If the writer of the sentence above takes precision to an extreme, it might result in the following:

> The incidence of high-speed, automobile accident related fatalities has decreased 75% and continued to remain at historical lows since the initial set of federal legislations requiring seat belt use were enacted in 1992.

This sentence is extremely precise, but it takes so long to achieve that precision that it suffers from a lack of clarity. Precise writing is about finding the right balance between information and flow. This is also an issue of *conciseness* (discussed in the next section).

The last thing to consider with precision is a word choice that's not only unclear or uninteresting, but also confusing or misleading. For example:

> The number of highway fatalities has become hugely lower since laws requiring seat belt use were enacted.

In this case, the reader might be confused by the word *hugely*. Huge means large, but here the writer uses *hugely* to describe something small. Though most readers can decipher this, doing so disconnects them from the flow of the writing and makes the writer's point less effective.

On the test, there can be questions asking for alternatives to the writer's word choice. In answering these questions, always consider the context and look for a balance between precision and flow.

Conciseness

"Less is more" is a good rule to follow when writing a sentence. Unfortunately, writers often include extra words and phrases that seem necessary at the time, but add nothing to the main idea. This confuses the reader and creates unnecessary repetition. Writing that lacks **conciseness** is usually guilty of excessive wordiness and redundant phrases. Here's an example containing both of these issues:

> When legislators decided to begin creating legislation making it mandatory for automobile drivers and passengers to make use of seat belts while in cars, a large number of them made those laws for reasons that were political reasons.

There are several empty or "fluff" words here that take up too much space. These can be eliminated while still maintaining the writer's meaning. For example:

- "decided to begin" could be shortened to "began"
- "making it mandatory for" could be shortened to "requiring"

- "make use of" could be shortened to "use"
- "a large number" could be shortened to "many"

In addition, there are several examples of redundancy that can be eliminated:

- "legislators decided to begin creating legislation" and "made those laws"
- "automobile drivers and passengers" and "while in cars"
- "reasons that were political reasons"

These changes are incorporated as follows:

> When legislators began requiring drivers and passengers to use seat belts, many of them did so for political reasons.

There are many examples of redundant phrases, such as "add an additional," "complete and total," "time schedule," and "transportation vehicle." If asked to identify a redundant phrase on the test, look for words that are close together with the same (or similar) meanings.

Style, Tone, and Mood

Style, tone, and mood are often thought to be the same thing. Though they're closely related, there are important differences to keep in mind. The easiest way to do this is to remember that style "creates and affects" tone and mood. More specifically, style is *how the writer uses words* to create the desired tone and mood for their writing.

Style

Style can include any number of technical writing choices, and some may have to be analyzed on the test. A few examples of style choices include:

- Sentence Construction: When presenting facts, does the writer use shorter sentences to create a quicker sense of the supporting evidence, or do they use longer sentences to elaborate and explain the information?

- Technical Language: Does the writer use jargon to demonstrate their expertise in the subject, or do they use ordinary language to help the reader understand things in simple terms?

- Formal Language: Does the writer refrain from using contractions such as *won't* or *can't* to create a more formal tone, or do they use a colloquial, conversational style to connect to the reader?

- Formatting: Does the writer use a series of shorter paragraphs to help the reader follow a line of argument, or do they use longer paragraphs to examine an issue in great detail and demonstrate their knowledge of the topic?

On the test, examine the writer's style and how their writing choices affect the way the passage comes across.

Tone

Tone refers to the writer's attitude toward the subject matter. Tone is usually explained in terms of a work of fiction. For example, the tone conveys how the writer feels about their characters and the situations in which they're involved. Nonfiction writing is sometimes thought to have no tone at all, but this is incorrect.

A lot of nonfiction writing has a neutral tone, which is an extremely important tone for the writer to take. A neutral tone demonstrates that the writer is presenting a topic impartially and letting the information speak for itself. On the other hand, nonfiction writing can be just as effective and appropriate if the tone isn't neutral. For instance, take the previous examples involving seat belt use. In them, the writer mostly chooses to retain a neutral tone when presenting information. If the writer would instead include their own personal experience of losing a friend or family member in a car accident, the tone would change dramatically. The tone would no longer be neutral. Now it would show that the writer has a personal stake in the content, allowing them to interpret the information in a different way. When analyzing tone, consider what the writer is trying to achieve in the passage, and how they *create* the tone using style.

Mood

Mood refers to the feelings and atmosphere that the writer's words create for the reader. Like tone, many nonfiction pieces can have a neutral mood. To return to the previous example, if the writer would choose to include information about a person they know being killed in a car accident, the passage would suddenly carry an emotional component that is absent in the previous examples. Depending on how they present the information, the writer can create a sad, angry, or even hopeful mood. When analyzing the mood, consider what the writer wants to accomplish and whether the best choice was made to achieve that end.

Consistency

Whatever style, tone, and mood the writer uses, good writing should remain **consistent** throughout. If the writer chooses to include the tragic, personal experience above, it would affect the style, tone, and mood of the entire piece. It would seem out of place for such an example to be used in the middle of a neutral, measured, and analytical piece. To adjust the rest of the piece, the writer needs to make additional choices to remain consistent. For example, the writer might decide to use the word *tragedy* in place of the more neutral *fatality*, or they could describe a series of car-related deaths as an *epidemic*. Adverbs and adjectives such as *devastating* or *horribly* could be included to maintain this consistent attitude toward the content. When analyzing writing, look for sudden shifts in style, tone, and mood, and consider whether the writer would be wiser to maintain the prevailing strategy.

Syntax

Syntax is the order of words in a sentence. While most of the writing on the test has proper syntax, there may be questions on ways to vary the syntax for effectiveness. One of the easiest writing mistakes to spot is **repetitive sentence structure**. For example:

> Seat belts are important. They save lives. People don't like to use them. We have to pass seat belt laws. Then more people will wear seat belts. More lives will be saved.

What's the first thing that comes to mind when reading this example? The short, choppy, and repetitive sentences! In fact, most people notice this syntax issue more than the content itself. By combining some sentences and changing the syntax of others, the writer can create a more effective writing passage:

> Seat belts are important because they save lives. Since people don't like to use seat belts, though, more laws requiring their usage need to be passed. Only then will more people wear them and only then will more lives be saved.

Many rhetorical devices can be used to vary syntax (more than can possibly be named here). These often have intimidating names like *anadiplosis*, *metastasis*, and *paremptosis*. The test questions don't ask for definitions of these tricky techniques, but they can ask how the writer plays with the words and what effect that has on the writing. For example, *anadiplosis* is when the last word (or phrase) from a sentence is used to begin the next sentence:

Cars are driven by people. People cause accidents. Accidents cost taxpayers money.

The test doesn't ask for this technique by name, but be prepared to recognize what the writer is doing and why they're using the technique in this situation. In this example, the writer is probably using *anadiplosis* to demonstrate causation.

Standard English Conventions

Sentence Structure

First, let's review the basic elements of sentences.

A **sentence** is a set of words that make up a grammatical unit. The words must have certain elements and be spoken or written in a specific order to constitute a complete sentence that makes sense.

1. A sentence must have a **subject** (a noun or noun phrase). The subject tells whom or what the sentence is addressing (i.e. what it is about).

2. A sentence must have an **action** or **state of being** (*a* verb). To reiterate: A verb forms the main part of the predicate of a sentence. This means that it explains what the noun is doing.

3. A sentence must convey a complete thought.

When examining writing, be mindful of grammar, structure, spelling, and patterns. Sentences can come in varying sizes and shapes; so, the point of grammatical correctness is not to stamp out creativity or diversity in writing. Rather, grammatical correctness ensures that writing will be enjoyable and clear. One of the most common methods for catching errors is to mouth the words as you read them. Many typos are fixed automatically by our brain, but mouthing the words often circumvents this instinct and helps one read what's actually on the page. Often, grammar errors are caught not by memorization of grammar rules but by the training of one's mind to know whether something *sounds* right or not.

Types of Sentences
There isn't an overabundance of absolutes in grammar, but here is one: every sentence in the English language falls into one of four categories.

- Declarative: a simple statement that ends with a period

   The price of milk per gallon is the same as the price of gasoline.

- Imperative: a command, instruction, or request that ends with a period

   Buy milk when you stop to fill up your car with gas.

- Interrogative: a question that ends with a question mark

     Will you buy the milk?

- Exclamatory: a statement or command that expresses emotions like anger, urgency, or surprise and ends with an exclamation mark

     Buy the milk now!

Declarative sentences are the most common type, probably because they are comprised of the most general content, without any of the bells and whistles that the other three types contain. They are, simply, declarations or statements of any degree of seriousness, importance, or information.

Imperative sentences often seem to be missing a subject. The subject is there, though; it is just not visible or audible because it is *implied*. Look at the imperative example sentence.

     Buy the milk when you fill up your car with gas.

You is the implied subject, the one to whom the command is issued. This is sometimes called *the understood you* because it is understood that *you* is the subject of the sentence.

Interrogative sentences—those that ask questions—are defined as such from the idea of the word *interrogation*, the action of questions being asked of suspects by investigators. Although that is serious business, interrogative sentences apply to all kinds of questions.

To exclaim is at the root of **exclamatory sentences**. These are made with strong emotions behind them. The only technical difference between a declarative or imperative sentence and an exclamatory one is the exclamation mark at the end. The example declarative and imperative sentences can both become an exclamatory one simply by putting an exclamation mark at the end of the sentences.

     The price of milk per gallon is the same as the price of gasoline!
     Buy milk when you stop to fill up your car with gas!

After all, someone might be really excited by the price of gas or milk, or they could be mad at the person that will be buying the milk! However, as stated before, exclamation marks in abundance defeat their own purpose! After a while, they begin to cause fatigue! When used only for their intended purpose, they can have their expected and desired effect.

Independent and Dependent Clauses

Independent and dependent clauses are strings of words that contain both a subject and a verb. An **independent clause** *can* stand alone as complete thought, but a **dependent clause** *cannot*. A dependent clause relies on other words to be a complete sentence.

     Independent clause: The keys are on the counter.
     Dependent clause: If the keys are on the counter

Notice that both clauses have a subject (*keys*) and a verb (*are*). The independent clause expresses a complete thought, but the word *if* at the beginning of the dependent clause makes it *dependent* on other words to be a complete thought.

     Independent clause: If the keys are on the counter, please give them to me.

This presents a complete sentence since it includes at least one verb and one subject and is a complete thought. In this case, the independent clause has two subjects (*keys* & an implied *you*) and two verbs (*are* & *give*).

> Independent clause: I went to the store.
> Dependent clause: Because we are out of milk,
>
> Complete Sentence: Because we are out of milk, I went to the store.
> Complete Sentence: I went to the store because we are out of milk.

Sentence Structures

A **simple sentence** has one independent clause.

> I am going to win.

A **compound sentence** has two independent clauses. A conjunction—*for, and, nor, but, or, yet, so*—links them together. Note that each of the independent clauses has a subject and a verb.

> I am going to win, but the odds are against me.

A **complex sentence** has one independent clause and one or more dependent clauses.

> I am going to win, even though I don't deserve it.

Even though I don't deserve it is a dependent clause. It does not stand on its own. Some conjunctions that link an independent and a dependent clause are *although*, *because*, *before*, *after*, *that*, *when*, *which*, and *while*.

A **compound-complex sentence** has at least three clauses, two of which are independent and at least one that is a dependent clause.

> While trying to dance, I tripped over my partner's feet, but I regained my balance quickly.

The dependent clause is *While trying to dance*.

Run-Ons and Fragments

Run-Ons

A common mistake in writing is the run-on sentence. A **run-on** is created when two or more independent clauses are joined without the use of a conjunction, a semicolon, a colon, or a dash. We don't want to use commas where periods belong. Here is an example of a run-on sentence:

> Making wedding cakes can take many hours I am very impatient, I want to see them completed right away.

There are a variety of ways to correct a run-on sentence. The method you choose will depend on the context of the sentence and how it fits with neighboring sentences:

> Making wedding cakes can take many hours. I am very impatient. I want to see them completed right away. (Use periods to create more than one sentence.)

Making wedding cakes can take many hours; I am very impatient—I want to see them completed right away. (Correct the sentence using a semicolon, colon, or dash.)

Making wedding cakes can take many hours, and I am very impatient and want to see them completed right away. (Correct the sentence using coordinating conjunctions.)

I am very impatient because I would rather see completed wedding cakes right away than wait for it to take many hours. (Correct the sentence by revising.)

Fragments

Remember that a complete sentence must have both a subject and a verb. Complete sentences consist of at least one independent clause. Incomplete sentences are called **sentence fragments**. A sentence fragment is a common error in writing. Sentence fragments can be independent clauses that start with subordinating words, such as *but, as, so that,* or *because,* or they could simply be missing a subject or verb.

You can correct a fragment error by adding the fragment to a nearby sentence or by adding or removing words to make it an independent clause. For example:

Dogs are my favorite animals. Because cats are too lazy. (Incorrect; the word because creates a sentence fragment)

Dogs are my favorite animals because cats are too lazy. (Correct; this is a dependent clause.)

Dogs are my favorite animals. Cats are too lazy. (Correct; this is a simple sentence.)

Subject and Predicate

Every complete sentence can be divided into two parts: the subject and the predicate.

Subjects: We need to have subjects in our sentences to tell us who or what the sentence describes. Subjects can be simple or complete, and they can be direct or indirect. There can also be compound subjects.

Simple subjects are the noun or nouns the sentence describes, without modifiers. The simple subject can come before or after the verb in the sentence:

The big brown <u>dog</u> is the calmest one.

Complete subjects are the subject together with all of its describing words or modifiers.

The <u>big brown dog</u> is the calmest one. (The complete subject is big brown dog.)

Direct subjects are subjects that appear in the text of the sentence, as in the example above. **Indirect subjects** are implied. The subject is "you," but the word *you* does not appear.

Indirect subjects are usually in imperative sentences that issue a command or order:

Feed the short skinny dog first. (The understood you is the subject.)

Watch out—he's really hungry! (The sentence warns you to watch out.)

Compound subjects occur when two or more nouns join together to form a plural subject.

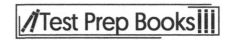

Carson and Emily make a great couple.

Predicates: Once we have identified the subject of the sentence, the rest of the sentence becomes the predicate. Predicates are formed by the verb, the direct object, and all words related to it.

We went to see the Cirque du' Soleil performance.

The gigantic green character was funnier than all the rest.

A **predicate nominative** renames the subject:

John is a carpenter.

A **predicate adjective** describes the subject:

Margaret is beautiful.

Direct objects are the nouns in the sentence that are receiving the action. Sentences don't necessarily need objects. Sentences only need a subject and a verb.

The clown brought the acrobat the hula-hoop. (What is getting brought? the hula-hoop)

Then he gave the trick pony a soapy bath. (What is being given? (a soapy bath)

Indirect objects are words that tell us to or for whom or what the action is being done. For there to be an indirect object, there first must always be a direct object.

The clown brought the acrobat the hula-hoop. (Who is getting the direct object? the hula-hoop)

Then he gave the trick pony a soapy bath. (What is getting the bath? a trick pony)

Phrases

A **phrase** is a group of words that go together but do not include both a subject and a verb. We use them to add information, explain something, or make the sentence easier for the reader to understand. Unlike clauses, phrases can never stand alone as their own sentence. They do not form complete thoughts. There are noun phrases, prepositional phrases, verbal phrases, appositive phrases, and absolute phrases. Here are some examples of phrases:

I know all the shortest routes.

Before the sequel, we wanted to watch the first movie. (introductory phrase)

The jumpers have hot cocoa to drink right away.

Complements

A **complement** completes the meaning of an expression. A complement can be a pronoun, noun, or adjective. A verb complement refers to the direct object or indirect object in the sentence. An object complement gives more information about the direct object:

The magician got the kids excited.

Kids is the direct object, and *excited* is the object complement.

A *subject complement* comes after a linking verb. It is typically an adjective or noun that gives more information about the subject:

> The king was noble and spared the thief's life.

Noble describes the *king* and follows the linking verb *was*.

Subject-Verb Agreement

The subject of a sentence and its verb must agree. The cornerstone rule of subject-verb agreement is that subject and verb must agree in number. Whether the subject is singular or plural, the verb must follow suit.

> Incorrect: The houses is new.
>
> Correct: The houses are new.
>
> Also Correct: The house is new.

In other words, a singular subject requires a singular verb; a plural subject requires a plural verb.

The words or phrases that come between the subject and verb do not alter this rule.

> Incorrect: The houses built of brick is new.
>
> Correct: The houses built of brick are new.
>
> Incorrect: The houses with the sturdy porches is new.
>
> Correct: The houses with the sturdy porches are new.

The subject will always follow the verb when a sentence begins with *here* or *there.* Identify these with care.

> Incorrect: Here *is* the *houses* with sturdy porches.
>
> Correct: Here *are* the *houses* with sturdy porches.

The subject in the sentences above is not *here*, it is *houses*. Remember, *here* and *there* are never subjects. Be careful that contractions such as *here's* or *there're* do not cause confusion!

Two subjects joined by *and* require a plural verb form, except when the two combine to make one thing:

> Incorrect: Garrett and Jonathan is over there.
>
> Correct: Garrett and Jonathan are over there.
>
> Incorrect: Spaghetti and meatballs are a delicious meal!
>
> Correct: Spaghetti and meatballs is a delicious meal!

In the example above, *spaghetti and meatballs* is a compound noun. However, *Garrett and Jonathan* is not a compound noun.

Two singular subjects joined by *or, either/or,* or *neither/nor* call for a singular verb form.

Incorrect: Butter or syrup are acceptable.

Correct: Butter or syrup is acceptable.

Plural subjects joined by *or, either/or*, or *neither/nor* are, indeed, plural.

The chairs or the boxes are being moved next.

If one subject is singular and the other is plural, the verb should agree with the closest noun.

Correct: The chair or the boxes are being moved next.

Correct: The chairs or the box is being moved next.

Some plurals of money, distance, and time call for a singular verb.

Incorrect: Three dollars *are* enough to buy that.

Correct: Three dollars *is* enough to buy that.

For words declaring degrees of quantity such as *many of, some of,* or *most of,* let the noun that follows *of* be the guide:

Incorrect: Many of the books is in the shelf.

Correct: Many of the books are in the shelf.

Incorrect: Most of the pie *are* on the table.

Correct: Most of the pie *is* on the table.

For indefinite pronouns like anybody or everybody, use singular verbs.

Everybody *is* going to the store.

However, the pronouns *few, many, several, all, some,* and *both* have their own rules and use plural forms.

Some *are* ready.

Some nouns like *crowd* and *congress* are called *collective nouns* and they require a singular verb form.

Congress *is* in session.

The news *is* over.

Books and movie titles, though, including plural nouns such as *Great Expectations*, also require a singular verb. Remember that only the subject affects the verb. While writing tricky subject-verb arrangements, say them aloud. Listen to them. Once the rules have been learned, one's ear will become sensitive to them, making it easier to pick out what's right and what's wrong.

Dangling and Misplaced Modifiers

A **modifier** is a word or phrase meant to describe or clarify another word in the sentence. When a sentence has a modifier but is missing the word it describes or clarifies, it's an error called a **dangling modifier**. We can fix the sentence by revising to include the word that is being modified.

Consider the following examples with the modifier underlined:

Incorrect: <u>Having walked five miles</u>, this bench will be the place to rest. (This implies that the bench walked the miles, not the person.)

Correct: <u>Having walked five miles</u>, Matt will rest on this bench. (*Having walked five miles* correctly modifies *Matt*, who did the walking.)

Incorrect: <u>Since midnight</u>, my dreams have been pleasant and comforting. (The adverb clause *since midnight* cannot modify the noun *dreams*.)

Correct: <u>Since midnight</u>, I have had pleasant and comforting dreams. (*Since midnight* modifies the verb have had, telling us when the dreams occurred.)

Sometimes the modifier is not located close enough to the word it modifies for the sentence to be clearly understood. In this case, we call the error a **misplaced modifier**. Here is an example with the modifier underlined.

Incorrect: We gave the hot cocoa to the children <u>that was filled with marshmallows</u>. (This sentence implies that the children are what are filled with marshmallows.)

Correct: We gave the hot cocoa <u>that was filled with marshmallows</u> to the children. (The cocoa is filled with marshmallows. The modifier is near the word it modifies.)

Parallel Structure in a Sentence

Parallel structure, also known as **parallelism**, refers to using the same grammatical form within a sentence. This is important in lists and for other components of sentences.

Incorrect: At the recital, the boys and girls were dancing, singing, and played musical instruments.
Correct: At the recital, the boys and girls were dancing, singing, and playing musical instruments.

Notice that in the second example, *played* is not in the same verb tense as the other verbs nor is it compatible with the helping verb *were*. To test for parallel structure in lists, try reading each item as if it were the only item in the list.

The boys and girls were dancing.
The boys and girls were singing.
The boys and girls were played musical instruments.

Suddenly, the error in the sentence becomes very clear. Here's another example:

> Incorrect: After the accident, I informed the police *that Mrs. Holmes backed* into my car, *that Mrs. Holmes got out* of her car to look at the damage, and *she was driving* off without leaving a note.

> Correct: After the accident, I informed the police *that Mrs. Holmes backed* into my car, *that Mrs. Holmes got out* of her car to look at the damage, and *that Mrs. Holmes drove off* without leaving a note.

> Correct: After the accident, I informed the police that Mrs. Holmes *backed* into my car, *got out* of her car to look at the damage, and *drove off* without leaving a note.

Note that there are two ways to fix the nonparallel structure of the first sentence. The key to parallelism is consistent structure.

Usage

Parts of Speech

Nouns

A **common noun** is a word that identifies any of a class of people, places, or things. Examples include numbers, objects, animals, feelings, concepts, qualities, and actions. *A, an,* or *the* usually precedes the common noun. These parts of speech are called *articles*. Here are some examples of sentences using nouns preceded by articles.

> *A* building is under construction.

> *The* girl would like to move to *the* city.

An **abstract noun** is an idea, state, or quality. It is something that can't be touched, such as happiness, courage, evil, or humor.

A **proper noun** (also called a **proper name**) is used for the specific name of an individual person, place, or organization. The first letter in a proper noun is capitalized. "My name is *Mary*." "I work for *Walmart*."

Nouns sometimes serve as adjectives (which themselves describe nouns), such as "hockey player" and "state government."

Pronouns

A word used in place of a noun is known as a **pronoun**. Pronouns are words like *I, mine, hers,* and *us*.

Pronouns can be split into different classifications (as shown below) which make them easier to learn; however, it's not important to memorize the classifications.

- **Personal pronouns:** refer to people

- **First person pronouns:** we, I, our, mine

- **Second person pronouns:** you, yours

- **Third person pronouns:** he, she, they, them, it

- **Possessive pronouns:** demonstrate ownership (mine, his, hers, its, ours, theirs, yours)

- **Interrogative pronouns:** ask questions (what, which, who, whom, whose)

- **Relative pronouns:** include the five interrogative pronouns and others that are relative (whoever, whomever, that, when, where)

- **Demonstrative pronouns:** replace something specific (this, that, those, these)

- **Reciprocal pronouns:** indicate something was done or given in return (each other, one another)

- **Indefinite pronouns:** have a nonspecific status (anybody, whoever, someone, everybody, somebody)

Indefinite pronouns such as *anybody, whoever, someone, everybody,* and *somebody* command a singular verb form, but others such as *all, none,* and *some* could require a singular or plural verb form.

Antecedents

An **antecedent** is the noun to which a pronoun refers; it needs to be written or spoken before the pronoun is used. For many pronouns, antecedents are imperative for clarity. In particular, a lot of the personal, possessive, and demonstrative pronouns need antecedents. Otherwise, it would be unclear who or what someone is referring to when they use a pronoun like *he* or *this.*

Pronoun reference means that the pronoun should refer clearly to one, clear, unmistakable noun (the antecedent).

Pronoun-antecedent agreement refers to the need for the antecedent and the corresponding pronoun to agree in gender, person, and number. Here are some examples:

The *kidneys* (plural antecedent) are part of the urinary system. *They* (plural pronoun) serve several roles."

The kidneys are part of the *urinary system* (singular antecedent). *It* (singular pronoun) is also known as the renal system.

Pronoun Cases

The **subjective pronouns** —*I, you, he/she/it, we, they,* and *who*—are the subjects of the sentence.

Example: *They* have a new house.

The **objective pronouns**—*me, you* (*singular*)*, him/her, us, them,* and *whom*—are used when something is being done for or given to someone; they are objects of the action.

 Example: The teacher has an apple for *us*.

The **possessive pronouns**—*mine, my, your, yours, his, hers, its, their, theirs, our,* and *ours*—are used to denote that something (or someone) belongs to someone (or something).

 Example: It's *their* chocolate cake.

 Even Better Example: It's *my* chocolate cake!

One of the greatest challenges and worst abuses of pronouns concerns *who* and *whom*. Just knowing the following rule can eliminate confusion. *Who* is a subjective-case pronoun used only as a subject or subject complement. *Whom* is only objective-case and, therefore, the object of the verb or preposition.

 Who is going to the concert?

 You are going to the concert with *whom*?

Hint: When using *who* or *whom*, think of whether someone would say *he* or *him*. If the answer is *he*, use *who*. If the answer is *him*, use *whom*. This trick is easy to remember because *he* and *who* both end in vowels, and *him* and *whom* both end in the letter *M*.

Many possessive pronouns sound like contractions. For example, many people get *it's* and *its* confused. The word *it's* is the contraction for *it is*. The word *its* without an apostrophe is the possessive form of *it*.

 I love that wooden desk. It's beautiful. (contraction)

 I love that wooden desk. Its glossy finish is beautiful. (possessive)

If you are not sure which version to use, replace *it's/its* with *it is* and see if that sounds correct. If so, use the contraction (*it's*). That trick also works for *who's/whose, you're/your,* and *they're/their*.

Adjectives
"The *extraordinary* brain is the *main* organ of the central nervous system." The adjective *extraordinary* describes the brain in a way that causes one to realize it is more exceptional than some of the other organs while the adjective *main* defines the brain's importance in its system.

An **adjective** is a word or phrase that names an attribute that describes or clarifies a noun or pronoun. This helps the reader visualize and understand the characteristics—size, shape, age, color, origin, etc.—of a person, place, or thing that otherwise might not be known. Adjectives breathe life, color, and depth into the subjects they define. Life would be *drab* and *colorless* without adjectives!

Adjectives often precede the nouns they describe.

 She drove her <u>new</u> car.

However, adjectives can also come later in the sentence.

 Her car is <u>new</u>.

Adjectives using the prefix *a–* can only be used after a verb.

> Correct: The dog was *alive* until the car ran up on the curb and hit him.

> Incorrect: The *alive* dog was hit by a car that ran up on the curb.

Other examples of this rule include *awake, ablaze, ajar, alike,* and *asleep.*

Other adjectives used after verbs concern states of health.

> The girl was finally *well* after a long bout of pneumonia.

> The boy was *fine* after the accident.

An adjective phrase is not a bunch of adjectives strung together, but a group of words that describes a noun or pronoun and, thus, functions as an adjective. Very happy is an adjective phrase; so are way too hungry and passionate about traveling.

Possessives

In grammar, *possessive nouns* show ownership, which was seen in previous examples like *mine, yours,* and *theirs.*

Singular nouns are generally made possessive with an apostrophe and an *s* (*'s*).

> My *uncle's* new car is silver.

> The *dog's* bowl is empty.

> *James's* ties are becoming outdated.

Plural nouns ending in *s* are generally made possessive by just adding an apostrophe (*'*):

> The pistachio nuts' saltiness is added during roasting. (The saltiness of pistachio nuts is added during roasting.)

> The students' achievement tests are difficult. (The achievement tests of the students are difficult.)

If the plural noun does not end in an *s* such as *women,* then it is made possessive by adding an *apostrophe s* (*'s*)—*women's.*

Indefinite possessive pronouns such as *nobody* or *someone* become possessive by adding an *apostrophe s*— *nobody's* or *someone's.*

Verbs

The **verb** is the part of speech that describes an action, state of being, or occurrence.

A verb forms the main part of a predicate of a sentence. This means that the verb explains what the noun (which will be discussed shortly) is doing. A simple example is *time flies.* The verb *flies* explains what the action of the noun, *time,* is doing. This example is a *main* verb.

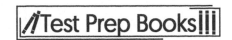

Helping (auxiliary) verbs are words like *have, do, be, can, may, should, must,* and *will.* "I *should* go to the store." Helping verbs assist main verbs in expressing tense, ability, possibility, permission, or obligation.

Particles are minor function words like *not, in, out, up,* or *down* that become part of the verb itself. "I might *not.*"

Participles are words formed from verbs that are often used to modify a noun, noun phrase, verb, or verb phrase.

> The *running* teenager collided with the cyclist.

Participles can also create compound verb forms.

> He is *speaking.*

Participial phrases are made up of the participle and modifiers, complements, or objects.

> *Crying for most of an hour*, the baby didn't seem to want to nap.

> *Having already taken this course*, the student was bored during class.

Verbs have five basic forms: the **base** form, the **-s** form, the **-ing** form, the **past** form, and the **past participle** form.

The past forms are either **regular** (*love/loved; hate/hated*) or **irregular** because they don't end by adding the common past tense suffix "-ed" (*go/went; fall/fell; set/set*).

Adverbs

Adverbs have more functions than adjectives because they modify or qualify verbs, adjectives, or other adverbs as well as word groups that express a relation of place, time, circumstance, or cause. Therefore, adverbs answer any of the following questions: *How, when, where, why, in what way, how often, how much, in what condition,* and/or *to what degree. How good looking is he? He is <u>very</u> handsome.*

Here are some examples of adverbs for different situations:

- how: quickly
- when: daily
- where: there
- in what way: easily
- how often: often
- how much: much
- in what condition: badly
- what degree: hardly

As one can see, for some reason, many adverbs end in *-ly.*

Adverbs do things like emphasize (*really, simply,* and *so*), amplify (*heartily, completely,* and *positively*), and tone down (*almost, somewhat,* and *mildly*).

Adverbs also come in phrases.

The dog ran as <u>though his life depended on it.</u>

Prepositions

Prepositions are connecting words and, while there are only about 150 of them, they are used more often than any other individual groups of words. They describe relationships between other words. They are placed before a noun or pronoun, forming a phrase that modifies another word in the sentence. **Prepositional phrases** begin with a preposition and end with a noun or pronoun, the **object of the preposition.** *A pristine lake is <u>near the store</u> and <u>behind the bank.</u>*

Some commonly used prepositions are *about, after, anti, around, as, at, behind, beside, by, for, from, in, into, of, off, on, to,* and *with.*

Complex prepositions, which also come before a noun or pronoun, consist of two or three words such as *according to, in regards to,* and *because of.*

Interjections

Interjections are words used to express emotion. Examples include *wow, ouch,* and *hooray.* Interjections are often separate from sentences; in those cases, the interjection is directly followed by an exclamation point. In other cases, the interjection is included in a sentence and followed by a comma. The punctuation plays a big role in the intensity of the emotion that the interjection is expressing. Using a comma or semicolon indicates less excitement than using an exclamation mark.

Conjunctions

Conjunctions are vital words that connect words, phrases, thoughts, and ideas. Conjunctions show relationships between components. There are two types:

Coordinating conjunctions are the primary class of conjunctions placed between words, phrases, clauses, and sentences that are of equal grammatical rank; the coordinating conjunctions are *for, and, nor, but, or, yet,* and *so.* A useful memorization trick is to remember that all the first letters of these conjunctions collectively spell the word fanboys.

I need to go shopping, *but* I must be careful to leave enough money in the bank.

She wore a black, red, *and* white shirt.

Subordinating conjunctions are the secondary class of conjunctions. They connect two unequal parts, one **main** (or **independent**) and the other **subordinate** (or **dependent**). I must go to the store *even though* I do not have enough money in the bank.

Because I read the review, I do not want to go to the movie.

Notice that the presence of subordinating conjunctions makes clauses dependent. *I read the review* is an independent clause, but *because* makes the clause dependent. Thus, it needs an independent clause to complete the sentence.

Shifts in Construction

It's been said several times already that *good writing must be consistent.* Another common writing mistake occurs when the writer unintentionally shifts verb tense, voice, or noun-pronoun agreement.

This shift can take place within a sentence, within a paragraph, or over the course of an entire piece of writing. On the test, questions may ask that this kind of error be identified. Here are some examples.

Shift in Verb Tense

Verb tense reflects when an action occurred or a state existed. For example, the tense known as **simple present** expresses something that is happening right now or that happens regularly:

She *works* in a hospital.

Present continuous tense expresses something in progress. It is formed by to be + verb + -ing.

Sorry, I can't go out right now. I *am doing* my homework.

Past tense is used to describe events that previously occurred. However, in conversational English, speakers often use present tense or a mix of past and present tense when relating past events because it gives the narrative a sense of immediacy. In formal written English, though, consistency in verb tense is necessary to avoid reader confusion.

I traveled to Europe last summer. As soon as I stepped off the plane, I feel like I'm in a movie! I'm surrounded by quaint cafes and impressive architecture.

The passage above abruptly switches from past tense—*traveled, stepped*—to present tense—*feel, am surrounded*.

I *traveled* to Europe last summer. As soon as I *stepped* off the plane, I *felt* like I was in a movie! I *was surrounded* by quaint cafes and impressive architecture.

All verbs are in past tense, so this passage now has consistent verb tense.

Shift in Voice

Sometimes the writer accidentally slips from active voice to passive voice in the middle of a sentence. This is a difficult mistake to catch because it's something people often do when speaking to one another. First, it's important to understand the difference between active and passive voice. Most sentences are written in **active voice**, which means that the noun is doing what the verb in the sentence says. For example:

Seat belts save lives.

Here, the noun (*seat belt*) is doing the saving. However, in **passive voice**, the verb is doing something to the noun:

Lives are saved.

In this case, the noun (*lives*) is the thing *being saved*. Passive voice is difficult for many people to identify and understand, but there's a simple (and memorable) way to check: simply add "by zombies" to the end of the verb and, if it makes sense, then the verb is written in passive voice. For example: "My car was wrecked…by zombies."   Also, in the above example, "Lives are saved…by zombies." If the zombie trick doesn't work, then the sentence is in active voice.

Here's what a shift in voice looks like in a sentence:

> When Amy buckled her seat belt, a satisfying click was heard.

The writer shifts from active voice in the beginning of the sentence to passive voice after the comma (remember, "a satisfying click was heard…by zombies"). To fix this mistake, the writer must remain in active voice throughout:

> When Amy buckled her seat belt, she heard a satisfying click.

This sentence is now grammatically correct, easier to read…and zombie free!

Shift in Noun-Pronoun Agreement

Pronouns are used to replace nouns so sentences don't have a lot of unnecessary repetition. This repetition can make a sentence seem awkward as in the following example:

> Seat belts are important because seat belts save lives, but seat belts can't do so unless seat belts are used.

Replacing some of the nouns (*seat belts*) with a pronoun (*they*) improves the flow of the sentence:

> Seat belts are important because they save lives, but they can't do so unless they are used.

A pronoun should agree in number (singular or plural) with the noun that precedes it. Another common writing error is the shift in **noun-pronoun agreement**. Here's an example:

> When people are getting in a car, he should always remember to buckle his seatbelt.

The first half of the sentence talks about a plural (*people*), while the second half refers to a singular person (*he* and *his*). These don't agree, so the sentence should be rewritten as:

> When people are getting in a car, they should always remember to buckle their seatbelt.

Split Infinitives

The **infinitive form** of a verb consists of "to + base verb"—e.g., to walk, to sleep, to approve. A **split infinitive** occurs when another word, usually an adverb, is placed between *to* and the verb:

> I decided *to simply walk* to work to get more exercise every day.

The infinitive *to walk* is split by the adverb *simply*.

> It was a mistake *to hastily approve* the project before conducting further preliminary research.

The infinitive *to approve* is split by *hastily*.

Although some grammarians still advise against split infinitives, this syntactic structure is common in both spoken and written English and is widely accepted in standard usage.

Pronouns

Pronoun Person

Pronoun person refers to the narrative voice the writer uses in a piece of writing. A great deal of nonfiction is written in third person, which uses pronouns like *he, she, it,* and *they* to convey meaning.

Occasionally a writer uses first person (*I, me, we*, etc.) or second person (*you*). Any choice of pronoun person can be appropriate for a particular situation, but the writer must remain consistent and logical.

Test questions may cover examining samples that should stay in a single pronoun person, be it first, second, or third. Look out for shifts between words like *you* and *I* or *he* and *they*.

Pronoun Clarity

Pronouns always refer back to a noun. However, as the writer composes longer, more complicated sentences, the reader may be unsure which noun the pronoun should replace. For example:

> An amendment was made to the bill, but now it has been voted down.

Was the amendment voted down or the entire bill? It's impossible to tell from this sentence. To correct this error, the writer needs to restate the appropriate noun rather than using a pronoun:

> An amendment was made to the bill, but now the bill has been voted down.

Pronouns in Combination

Writers often make mistakes when choosing pronouns to use in combination with other nouns. The most common mistakes are found in sentences like this:

> Please join Senator Wilson and I at the event tomorrow.

Notice anything wrong? Though many people think the sentence sounds perfectly fine, the use of the pronoun *I* is actually incorrect. To double-check this, take the other person out of the sentence:

> Please join I at the event tomorrow.

Now the sentence is obviously incorrect, as it should read, "Please join *me* at the event tomorrow." Thus, the first sentence should replace *I* with *me*:

> Please join Senator Wilson and me at the event tomorrow.

For many people, this sounds wrong because they're used to hearing and saying it incorrectly. Take extra care when answering this kind of question and follow the double-checking procedure.

Agreement

In English writing, certain words connect to other words. People often learn these connections (or **agreements**) as young children and use the correct combinations without a second thought. However, the questions on the test dealing with agreement probably aren't simple ones.

Subject-Verb Agreement

Which of the following sentences is correct?

> A large crowd of protesters was on hand.

> A large crowd of protesters were on hand.

Many people would say the second sentence is correct, but they'd be wrong. However, they probably wouldn't be alone. Most people just look at two words: *protesters were*. Together they make sense. They sound right. The problem is that the verb *were* doesn't refer to the word *protesters*. Here, the word *protesters* is part of a prepositional phrase that clarifies the actual subject of the sentence (*crowd*). Take the phrase "of protesters" away and re-examine the sentences:

A large crowd was on hand.

A large crowd were on hand.

Without the prepositional phrase to separate the subject and verb, the answer is obvious. The first sentence is correct. On the test, look for confusing prepositional phrases when answering questions about subject-verb agreement. Take the phrase away, and then recheck the sentence.

Noun Agreement

Nouns that refer to other nouns must also match in number. Take the following example:

John and Emily both served as an intern for Senator Wilson.

Two people are involved in this sentence: John and Emily. Therefore, the word *intern* should be plural to match. Here is how the sentence should read:

John and Emily both served as interns for Senator Wilson.

Frequently Confused Words

There are a handful of words in the English language that writers often confuse with other words because they sound similar or identical. Errors involving these words are hard to spot because they *sound* right even when they're wrong. Also, because these mistakes are so pervasive, many people think they're correct. Here are a few examples that may be encountered on the test:

They're vs. Their vs. There

This set of words is probably the all-time winner of misuse. The word *they're* is a contraction of "they are." Remember that contractions combine two words, using an apostrophe to replace any eliminated letters. If a question asks whether the writer is using the word *they're* correctly, change the word to "they are" and reread the sentence. Look at the following example:

Legislators can be proud of they're work on this issue.

This sentence *sounds* correct, but replace the contraction *they're* with "they are" to see what happens:

Legislators can be proud of they are work on this issue.

The result doesn't make sense, which shows that it's an incorrect use of the word *they're*. Did the writer mean to use the word *their* instead? The word *their* indicates possession because it shows that something *belongs* to something else. Now put the word *their* into the sentence:

Legislators can be proud of their work on this issue.

To check the answer, find the word that comes right after the word *their* (which in this case is *work*). Pose this question: whose *work* is it? If the question can be answered in the sentence, then the word signifies possession. In the sentence above, it's the legislators' work. Therefore, the writer is using the word *their* correctly.

If the words *they're* and *their* don't make sense in the sentence, then the correct word is almost always *there*. The word *there* can be used in many different ways, so it's easy to remember to use it when *they're* and *their* don't work. Now test these methods with the following sentences:

Their going to have a hard time passing these laws.

Enforcement officials will have there hands full.

They're are many issues to consider when discussing car safety.

In the first sentence, asking the question "Whose going is it?" doesn't make sense. Thus the word *their* is wrong. However, when replaced with the conjunction *they're* (or *they are*), the sentence works. Thus the correct word for the first sentence should be *they're*.

In the second sentence, ask this question: "Whose hands are full?" The answer (*enforcement officials*) is correct in the sentence. Therefore, the word *their* should replace *there* in this sentence.

In the third sentence, changing the word *they're* to "they are" ("They are are many issues") doesn't make sense. Ask this question: "Whose are is it?" This makes even less sense, since neither of the words *they're* or *their* makes sense. Therefore, the correct word must be *there*.

Who's vs. Whose

Who's is a contraction of "who is" while the word *whose* indicates possession. Look at the following sentence:

Who's job is it to protect America's drivers?

The easiest way to check for correct usage is to replace the word *who's* with "who is" and see if the sentence makes sense:

Who is job is it to protect America's drivers?

By changing the contraction to "Who is" the sentence no longer makes sense. Therefore, the correct word must be *whose*.

Your vs. You're

The word *your* indicates possession, while *you're* is a contraction for "you are." Look at the following example:

Your going to have to write your congressman if you want to see action.

Again, the easiest way to check correct usage is to replace the word *Your* with "You are" and see if the sentence still makes sense.

You are going to have to write your congressman if you want to see action.

By replacing Your with "You are," the sentence still makes sense. Thus, in this case, the writer should have used "You're."

Its vs. It's

Its is a word that indicates possession, while the word *it's* is a contraction of "it is." Once again, the easiest way to check for correct usage is to replace the word with "it is" and see if the sentence makes sense. Look at the following sentence:

It's going to take a lot of work to pass this law.

Replacing *it's* with "it is" results in this: "It is going to take a lot of work to pass this law." This makes sense, so the contraction (*it's*) is correct. Now look at another example:

The car company will have to redesign it's vehicles.

Replacing *it's* with "it is" results in this: "The car company will have to redesign it is vehicles." This sentence doesn't make sense, so the contraction (*it's*) is incorrect.

Than vs. Then

Than is used in sentences that involve comparisons, while *then* is used to indicate an order of events. Consider the following sentence:

Japan has more traffic fatalities than the U.S.

The use of the word *than* is correct because it compares Japan to the U.S. Now look at another example:

Laws must be passed, and then we'll see a change in behavior.

Here the use of the word *then* is correct because one thing happens after the other.

Affect vs. Effect

Affect is a verb that means to change something, while *effect* is a noun that indicates such a change. Look at the following sentence:

There are thousands of people affected by the new law.

This sentence is correct because *affected* is a verb that tells what's happening. Now look at this sentence:

The law will have a dramatic effect.

This sentence is also correct because *effect* is a noun and the thing that happens.

Note that a noun version of *affect* is occasionally used. It means "emotion" or "desire," usually in a psychological sense.

Two vs. Too vs. To

Two is the number (2). *Too* refers to an amount of something, or it can mean *also*. *To* is used for everything else. Look at the following sentence:

Two senators still haven't signed the bill.

This is correct because there are *two* (2) senators. Here's another example:

There are too many questions about this issue.

In this sentence, the word *too* refers to an amount ("too many questions"). Now here's another example:

Senator Wilson is supporting this legislation, too.

In this sentence, the word *also* can be substituted for the word *too*, so it's also correct. Finally, one last example:

I look forward to signing this bill into law.

In this sentence, the tests for *two* and *too* don't work. Thus the word *to* fits the bill!

Other Common Writing Confusions

In addition to all of the above, there are other words that writers often misuse. This doesn't happen because the words sound alike, but because the writer is not aware of the proper way to use them.

Logical Comparison

Writers often make comparisons in their writing. However, it's easy to make mistakes in sentences that involve comparisons, and those mistakes are difficult to spot. Try to find the error in the following sentence:

Senator Wilson's proposed seat belt legislation was similar to Senator Abernathy.

Can't find it? First, ask what two things are actually being compared. It seems like the writer *wants* to compare two different types of legislation, but the sentence actually compares legislation ("Senator Wilson's proposed seat belt legislation") to a person ("Senator Abernathy"). This is a strange and illogical comparison to make.

So how can the writer correct this mistake? The answer is to make sure that the second half of the sentence logically refers back to the first half. The most obvious way to do this is to repeat words:

Senator Wilson's proposed seat belt legislation was similar to Senator Abernathy's seat belt legislation.

Now the sentence is logically correct, but it's a little wordy and awkward. A better solution is to eliminate the word-for-word repetition by using suitable replacement words:

Senator Wilson's proposed seat belt legislation was similar to that of Senator Abernathy.

Senator Wilson's proposed seat belt legislation was similar to the bill offered by Senator Abernathy.

Here's another similar example:

More lives in the U.S. are saved by seat belts than Japan.

The writer probably means to compare lives saved by seat belts in the U.S. to lives saved by seat belts in Japan. Unfortunately, the sentence's meaning is garbled by an illogical comparison, and instead refers to U.S. lives saved *by Japan* rather than *in Japan*. To resolve this issue, first repeat the words and phrases needed to make an identical comparison:

More lives in the U.S. are saved by seat belts than lives in Japan are saved by seat belts.

Then, use a replacement word to clean up the repetitive text:

> More lives in the U.S. are saved by seat belts than in Japan.

Capitalization

- Capitalize the first word in a sentence and the first word in a quotation:

    The realtor showed them the house.

    Robert asked, "When can we get together for dinner again?"

- Capitalize proper nouns and words derived from them:

    We are visiting Germany in a few weeks.

    We will stay with our German relatives on our trip.

- Capitalize days of the week, months of the year, and holidays:

    The book club meets the last Thursday of every month.

    The baby is due in June.

    I decided to throw a Halloween party this year.

- Capitalize the main words in titles (referred to as *title case*), but not the articles, conjunctions, or prepositions:

    A Raisin in the Sun

    To Kill a Mockingbird

- Capitalize directional words that are used as names, but not when referencing a direction:

    The North won the Civil War.

    After making a left, go north on Rt. 476.

    She grew up on the West Coast.

    The winds came in from the west.

- Capitalize titles that go with names:

    Mrs. McFadden   Sir Alec Guinness   Lt. Madeline Suarez

- Capitalize familial relationships when referring to a *specific* person:

    I worked for my Uncle Steven last summer.

    Did you work for your uncle last summer?

Idiomatic Usage

A **figure of speech** (sometimes called an **idiom**) is a rhetorical device. It's a phrase that is not intended to be taken literally.

When the writer uses a figure of speech, their intention must be clear if it's to be used effectively. Some phrases can be interpreted in a number of ways, causing confusion for the reader. Look for clues to the writer's true intention to determine the best replacement. Likewise, some figures of speech may seem out of place in a more formal piece of writing. To show this, here is another example involving seat belts:

> Seat belts save more lives than any other automobile safety feature. Many studies show that airbags save lives as well, however not all cars have airbags. For example, some older cars don't. In addition, air bags aren't entirely reliable. For example, studies show that in 15 percent of accidents, airbags don't deploy as designed, but, on the other hand, seat belt malfunctions happen once in a blue moon.

Most people know that "once in a blue moon" refers to something that rarely happens. However, because the rest of the paragraph is straightforward and direct, using this figurative phrase distracts the reader. In this example, the earlier version is much more effective.

Now it's important to take a moment and review the meaning of the word *literally*. This is because it's one of the most misunderstood and misused words in the English language. *Literally* means that something is exactly what it says it is, and there can be no interpretation or exaggeration. Unfortunately, *literally* is often used for emphasis as in the following example:

> This morning, I literally couldn't get out of bed.

This sentence meant to say that the person was extremely tired and wasn't able to get up. However, the sentence can't *literally* be true unless that person was tied down to the bed, paralyzed, or affected by a strange situation that the writer (most likely) didn't intend. Here's another example:

> I literally died laughing.

The writer tried to say that something was very funny. However, unless they're writing this from beyond the grave, it can't *literally* be true.

Note that this doesn't mean that writers can't use figures of speech. The colorful use of language and idioms make writing more interesting and draw in the reader. However, for these kinds of expressions to be used correctly, they cannot include the word *literally*.

Punctuation

Commas

A **comma** (,) is the punctuation mark that signifies a pause—breath—between parts of a sentence. It denotes a break of flow. As with so many aspects of writing structure, authors will benefit by reading their writing aloud or mouthing the words. This can be particularly helpful if one is uncertain about whether the comma is needed.

In a complex sentence—one that contains a subordinate (dependent) clause or clauses—the use of a comma is dictated by where the subordinate clause is located. If the subordinate clause is located before the main clause, a comma is needed between the two clauses.

I will not pay for the steak, *because I don't have that much money.*

Generally, if the subordinate clause is placed after the main clause, no punctuation is needed.

I did well on my exam because I studied two hours the night before.

Notice how the last clause is dependent because it requires the earlier independent clauses to make sense.

Use a comma on both sides of an interrupting phrase.

I will pay for the ice cream, *chocolate and vanilla*, and then will eat it all myself.

The words forming the phrase in italics are nonessential (extra) information. To determine if a phrase is nonessential, try reading the sentence without the phrase and see if it's still coherent.

A comma is not necessary in this next sentence because no interruption—nonessential or extra information—has occurred. Read sentences aloud when uncertain.

I will pay for his chocolate and vanilla ice cream and then will eat it all myself.

If the nonessential phrase comes at the beginning of a sentence, a comma should only go at the end of the phrase. If the phrase comes at the end of a sentence, a comma should only go at the beginning of the phrase.

Other types of interruptions include the following:

- interjections: Oh no, I am not going.
- abbreviations: Barry Potter, M.D., specializes in heart disorders.
- direct addresses: Yes, Claudia, I am tired and going to bed.
- parenthetical phrases: His wife, lovely as she was, was not helpful.
- transitional phrases: Also, it is not possible.

The second comma in the following sentence is called an Oxford comma.

I will pay for ice cream, syrup, and pop.

It is a comma used after the second-to-last item in a series of three or more items. It comes before the word *or* or *and*. Not everyone uses the Oxford comma; it is optional, but many believe it is needed. The comma functions as a tool to reduce confusion in writing. So, if omitting the Oxford comma would cause confusion, then it's best to include it.

Commas are used in math to mark the place of thousands in numerals, breaking them up so they are easier to read. Other uses for commas are in dates (*March 19, 2016*), letter greetings (*Dear Sally,*), and in between cities and states (*Louisville, KY*).

Apostrophes

This punctuation mark, the apostrophe ('), is a versatile little mark. It has a few different functions:

- Quotes: Apostrophes are used when a second quote is needed within a quote.

- In my letter to my friend, I wrote, "The girl had to get a new purse, and guess what Mary did? She said, 'I'd like to go with you to the store.' I knew Mary would buy it for her."

- Contractions: Another use for an apostrophe in the quote above is a contraction. *I'd* is used for *I would.*

  The basic rule for making *contractions* is one area of spelling that is pretty straightforward: combine the two words by inserting an apostrophe (') in the space where a letter is omitted. For example, to combine *you* and *are*, drop the *a* and put the apostrophe in its place: *you're.*

  he + is = he's

  you + all = y'all (informal but often misspelled)

- Possession: An apostrophe followed by the letter *s* shows possession (*Mary's* purse). If the possessive word is plural, the apostrophe generally just follows the word.

- The trees' leaves are all over the ground.

Ellipses

An **ellipsis** (…) consists of three handy little dots that can speak volumes on behalf of irrelevant material. Writers use them in place of words, lines, phrases, list content, or paragraphs that might just as easily have been omitted from a passage of writing. This can be done to save space or to focus only on the specifically relevant material.

  Exercise is good for some unexpected reasons. Watkins writes, "Exercise has many benefits such as…reducing cancer risk."

In the example above, the ellipsis takes the place of the other benefits of exercise that are more expected.

The ellipsis may also be used to show a pause in sentence flow.

  "I'm wondering…how this could happen," Dylan said in a soft voice.

Semicolons

The **semicolon** (;) might be described as a heavy-handed comma. Take a look at these two examples:

  I will pay for the ice cream, but I will not pay for the steak.
  I will pay for the ice cream; I will not pay for the steak.

What's the difference? The first example has a comma and a conjunction separating the two independent clauses. The second example does not have a conjunction, but there are two independent clauses in the sentence, so something more than a comma is required. In this case, a semicolon is used.

Test Prep Books

Two independent clauses can only be joined in a sentence by either a comma and conjunction or a semicolon. If one of those tools is not used, the sentence will be a run-on. Remember that while the clauses are independent, they need to be closely related in order to be contained in one sentence.

Another use for the semicolon is to separate items in a list when the items themselves require commas.

> The family lived in Phoenix, Arizona; Oklahoma City, Oklahoma; and Raleigh, North Carolina.

Colons

Colons (:) have many miscellaneous functions. Colons can be used to precede further information or a list. In these cases, a colon should only follow an independent clause.

> Humans take in sensory information through five basic senses: sight, hearing, smell, touch, and taste.

The meal includes the following components:

- Caesar salad
- spaghetti
- garlic bread
- cake

The family got what they needed: a reliable vehicle.

While a comma is more common, a colon can also proceed a formal quotation.

> He said to the crowd: "Let's begin!"

The colon is used after the greeting in a formal letter.

> Dear Sir:
> To Whom It May Concern:

In the writing of time, the colon separates the minutes from the hour (*4:45 p.m.*). The colon can also be used to indicate a ratio between two numbers (*50:1*).

Hyphens

The **hyphen** (-) is a little hash mark that can be used to join words to show that they are linked.

Hyphenate two words that work together as a single adjective (a compound adjective).

> honey-covered biscuits

Some words always require hyphens, even if not serving as an adjective.

> merry-go-round

Hyphens always go after certain prefixes like *anti-* & *all-*.

Hyphens should also be used when the absence of the hyphen would cause a strange vowel combination (*semi-engineer*) or confusion. For example, *re-collect* should be used to describe something being gathered twice rather than being written as *recollect*, which means to remember.

Parentheses and Dashes

Parentheses are half-round brackets that look like this: ( ). They set off a word, phrase, or sentence that is an afterthought, explanation, or side note relevant to the surrounding text but not essential. A pair of commas is often used to set off this sort of information, but parentheses are generally used for information that would not fit well within a sentence or that the writer deems not important enough to be structurally part of the sentence.

> The picture of the heart (see above) shows the major parts you should memorize.
> Mount Everest is one of three mountains in the world that are over 28,000 feet high (K2 and Kanchenjunga are the other two).

See how the sentences above are complete without the parenthetical statements? In the first example, *see above* would not have fit well within the flow of the sentence. The second parenthetical statement could have been a separate sentence, but the writer deemed the information not pertinent to the topic.

The **em-dash** (—) is a mark longer than a hyphen used as a punctuation mark in sentences and to set apart a relevant thought. Even after plucking out the line separated by the dash marks, the sentence will be intact and make sense.

> Looking out the airplane window at the landmarks—Lake Clarke, Thompson Community College, and the bridge—she couldn't help but feel excited to be home.

The dashes use is similar to that of parentheses or a pair of commas. So, what's the difference? Many believe that using dashes makes the clause within them stand out while using parentheses is subtler. It's advised to not use dashes when commas could be used instead.

Quotation Marks

Here are some instances where *quotation marks* should be used:

- Dialogue for characters in narratives. When characters speak, the first word should always be capitalized, and the punctuation goes inside the quotes. For example:

  > Janie said, "The tree fell on my car during the hurricane."

- Around titles of songs, short stories, essays, and chapter in books
- To emphasize a certain word
- To refer to a word as the word itself

Capitalization Rules

Here's a non-exhaustive list of things that should be capitalized.

- The first word of every sentence
- The first word of every line of poetry
- The first letter of proper nouns (World War II)
- Holidays (Valentine's Day)
- The days of the week and months of the year (Tuesday, March)

- The first word, last word, and all major words in the titles of books, movies, songs, and other creative works (In the novel, *To Kill a Mockingbird*, note that *a* is lowercase since it's not a major word, but *to* is capitalized since it's the first word of the title.)
- Titles when preceding a proper noun (President Roberto Gonzales, Aunt Judy)

When simply using a word such as president or secretary, though, the word is not capitalized.

Officers of the new business must include a *president* and *treasurer*.

Seasons—spring, fall, etc.—are not capitalized.

North, *south*, *east*, and *west* are capitalized when referring to regions but are not when being used for directions. In general, if it's preceded by *the* it should be capitalized.

I'm from the South.
I drove south.

Homonyms

Homonyms are words that sound the same but are spelled differently, and they have different meanings. There are several common homonyms that give writers trouble.

There, They're, and *Their*
The word *there* can be used as an adverb, adjective, or pronoun:

There are ten children on the swim team this summer.

I put my book over *there*, but now I can't find it.

The word *they're* is a contraction of the words *they* and *are*:

They're flying in from Texas on Tuesday.

The word *their* is a possessive pronoun:

I store *their* winter clothes in the attic.

Its and *It's*
Its is a possessive pronoun:

The cat licked *its* injured paw.

It's is the contraction for the words *it* and *is*:

It's unbelievable how many people opted not to vote in the last election.

Your and You're
Your is a possessive pronoun:

Can I borrow *your* lawnmower this weekend?

You're is a contraction for the words *you* and *are*:

You're about to embark on a fantastic journey.

To, Too, and *Two*

To is an adverb or a preposition used to show direction, relationship, or purpose:

We are going *to* New York.

They are going *to* see a show.

Too is an adverb that means more than enough, also, and very:

You have had *too* much candy.

We are on vacation that week, *too*.

Two is the written-out form of the numeral 2:

Two of the shirts didn't fit, so I will have to return them.

New and *Knew*

New is an adjective that means recent:

There's a *new* customer on the phone.

Knew is the past tense of the verb *know*:

I *knew* you'd have fun on this ride.

Affect and *Effect*

Affect and *effect* are complicated because they are used as both nouns and verbs, have similar meanings, and are pronounced the same.

	Affect	**Effect**
Noun Definition	emotional state	result
Noun Example	The patient's affect was flat.	The effects of smoking are well documented.
Verb Definition	to influence	to bring about
Verb Example	The pollen count affects my allergies.	The new candidate hopes to effect change.

Independent and Dependent Clauses

Independent and *dependent* clauses are strings of words that contain both a subject and a verb. An independent clause *can* stand alone as complete thought, but a dependent clause *cannot*. A dependent clause relies on other words to be a complete sentence.

Independent clause: The keys are on the counter.
Dependent clause: If the keys are on the counter

Notice that both clauses have a subject (*keys*) and a verb (*are*). The independent clause expresses a complete thought, but the word *if* at the beginning of the dependent clause makes it *dependent* on other words to be a complete thought.

Independent clause: If the keys are on the counter, please give them to me.

This presents a complete sentence since it includes at least one verb and one subject and is a complete thought. In this case, the independent clause has two subjects (*keys* & an implied *you*) and two verbs (*are* & *give*).

Independent clause: I went to the store.
Dependent clause: Because we are out of milk,

Complete Sentence: Because we are out of milk, I went to the store.
Complete Sentence: I went to the store because we are out of milk.

Phrases

A *phrase* is a group of words that do not make a complete thought or a clause. They are parts of sentences or clauses. Phrases can be used as nouns, adjectives, or adverbs. A phrase does not contain both a subject and a verb.

Prepositional Phrases

A *prepositional phrase* shows the relationship between a word in the sentence and the object of the preposition. The object of the preposition is a noun that follows the preposition.

The orange pillows are on the couch.

On is the preposition, and *couch* is the object of the preposition.

She brought her friend with the nice car.

With is the preposition, and *car* is the object of the preposition. Here are some common prepositions:

about	as	at	after
by	for	from	in
of	on	to	with

Verbals and Verbal Phrases

Verbals are forms of verbs that act as other parts of speech. They can be used as nouns, adjectives, or adverbs. Though they are verb forms, they are not to be used as the verb in the sentence. A word group that is based on a verbal is considered a *verbal phrase*. There are three major types of verbals: *participles*, *gerunds*, and *infinitives*.

Participles are verbals that act as adjectives. The present participle ends in –*ing*, and the past participle ends in –*d*, -*ed*, -*n*, or-*t*.

Verb	Present Participle	Past Participle
walk	walking	walked
share	sharing	shared

Participial phrases are made up of the participle and modifiers, complements, or objects.

> Crying for most of an hour, the baby didn't seem to want to nap.

> Having already taken this course, the student was bored during class.

> *Crying for most of an hour* and *Having already taken this course* are the participial phrases.

Gerunds are verbals that are used as nouns and end in *–ing*. A gerund can be the subject or object of the sentence like a noun. Note that a present participle can also end in *–ing*, so it is important to distinguish between the two. The gerund is used as a noun, while the participle is used as an adjective.

> Swimming is my favorite sport.

> I wish I were sleeping.

A *gerund phrase* includes the gerund and any modifiers or complements, direct objects, indirect objects, or pronouns.

> Cleaning the house is my least favorite weekend activity.

Cleaning the house is the gerund phrase acting as the subject of the sentence.

> The most important goal this year is raising money for charity.

Raising money for charity is the gerund phrase acting as the direct object.

> The police accused the woman of stealing the car.

The *gerund* phrase *stealing the car* is the object of the preposition in this sentence.

An *infinitive* is a verbal made up of the word to and a verb. Infinitives can be used as nouns, adjectives, or adverbs.

> Examples: To eat, to jump, to swim, to lie, to call, to work

An *infinitive phrase* is made up of the infinitive plus any complements or modifiers. The infinitive phrase *to wait* is used as the subject in this sentence:

> To wait was not what I had in mind.

The infinitive phrase *to sing* is used as the subject complement in this sentence:

> Her dream is to sing.

The infinitive phrase *to grow* is used as an adverb in this sentence:

> Children must eat to grow.

Appositive Phrases

An *appositive* is a noun or noun phrase that renames a noun that comes immediately before it in the sentence. An appositive can be a single word or several words. These phrases can be *essential* or *nonessential*. An essential appositive phrase is necessary to the meaning of the sentence and a nonessential appositive phrase is not. It is important to be able to distinguish these for purposes of comma use.

Essential: My sister Christina works at a school.

Naming which sister is essential to the meaning of the sentence, so no commas are needed.

Nonessential: My sister, who is a teacher, is coming over for dinner tonight.

Who is a teacher is not essential to the meaning of the sentence, so commas are required.

Absolute Phrases

An *absolute phrase* modifies a noun without using a conjunction. It is not the subject of the sentence and is not a complete thought on its own. Absolute phrases are set off from the independent clause with a comma.

Arms outstretched, she yelled at the sky.

All things considered, this has been a great day.

The Four Types of Sentence Structures

A *simple sentence* has one independent clause.

I am going to win.

A *compound sentence* has two independent clauses. A conjunction—*for, and, nor, but, or, yet, so*—links them together. Note that each of the independent clauses has a subject and a verb.

I am going to win, but the odds are against me.

A *complex sentence* has one independent clause and one or more dependent clauses.

I am going to win, even though I don't deserve it.

Even though I don't deserve it is a dependent clause. It does not stand on its own. Some conjunctions that link an independent and a dependent clause are *although, because, before, after, that, when, which,* and *while.*

A *compound-complex sentence* has at least three clauses, two of which are independent and at least one that is a dependent clause.

While trying to dance, I tripped over my partner's feet, but I regained my balance quickly.

The dependent clause is *While trying to dance.*

Sentence Fragments

A *sentence fragment* is an incomplete sentence. An independent clause is made up of a subject and a predicate, and both are needed to make a complete sentence.

Sentence fragments often begin with relative pronouns (when, which), subordinating conjunctions (because, although) or gerunds (trying, being, seeing). They might be missing the subject or the predicate.

The most common type of fragment is the isolated dependent clause, which can be corrected by joining it to the independent clause that appears before or after the fragment:

Fragment: While the cookies baked.

Correction: While the cookies baked, we played cards. (We played cards while the cookies baked.)

Run-on Sentences

A *run-on sentence* is created when two independent clauses (complete thoughts) are joined without correct punctuation or a conjunction. Run-on sentences can be corrected in the following ways:

- Join the independent clauses with a comma and coordinating conjunction.

    Run-on: We forgot to return the library books we had to pay a fine.

    Correction: We forgot to return the library books, so we had to pay a fine.

- Join the independent clauses with a semicolon, dash, or colon when the clauses are closely related in meaning.

    Run-on: I had a salad for lunch every day this week I feel healthier already.

    Correction: I had a salad for lunch every day this week; I feel healthier already.

- Join the independent clauses with a *semicolon and a conjunctive adverb.*

    Run-on: We arrived at the animal shelter on time however the dog had already been adopted.

    Correction: We arrived at the animal shelter on time; however, the dog had already been adopted.

- Separate the independent clauses into two sentences *with a period.*

    Run-on: He tapes his favorite television show he never misses an episode.

    Correction: He tapes his favorite television show. He never misses an episode.

- *Rearrange the wording* of the sentence to create an independent clause and a dependent clause.

> Run-on: My wedding date is coming up I am getting more excited to walk down the aisle.

> Correction: As my wedding date approaches, I am getting more excited to walk down the aisle.

Dangling and Misplaced Modifiers

A *modifier* is a phrase that describes, alters, limits, or gives more information about a word in the sentence. The two most common issues are dangling and misplaced modifiers.

A *dangling modifier* is created when the phrase modifies a word that is not clearly stated in the sentence.

> Dangling modifier: Having finished dinner, the dishes were cleared from the table.

> Correction: Having finished dinner, Amy cleared the dishes from the table.

In the first sentence, *having finished dinner* appears to modify *the dishes*, which obviously can't finish dinner. The second sentence adds the subject *Amy*, to make it clear who has finished dinner.

> Dangling modifier: Hoping to improve test scores, all new books were ordered for the school.

> Correction: Hoping to improve test scores, administrators ordered all new books for the school.

> Without the subject *administrators*, it appears the books are hoping to improve test scores, which doesn't make sense.

Misplaced modifiers are placed incorrectly in the sentence, which can cause confusion. Compare these examples:

> Misplaced modifier: Rory purchased a new flat screen television and placed it on the wall above the fireplace, with all the bells and whistles.

> Revised: Rory purchased a new flat screen television, with all the bells and whistles, and placed it on the wall above the fireplace.

The bells and whistles should modify the television, not the fireplace.

> Misplaced modifier: The delivery driver arrived late with the pizza, who was usually on time.

> Revised: The delivery driver, who usually was on time, arrived late with the pizza.

This suggests that the delivery driver was usually on time, instead of the pizza.

> Misplaced modifier: We saw a family of ducks on the way to church.

> Revised: On the way to church, we saw a family of ducks.

> The misplaced modifier, here, suggests the *ducks* were on their way to church, instead of the pronoun *we*.

Split Infinitives

An infinitive is made up of the word *to* and a verb, such as: to run, to jump, to ask. A *split infinitive* is created when a word comes between *to* and the verb.

> Split infinitive: To quickly run

> Correction: To run quickly

> Split infinitive: To quietly ask

> Correction: To ask quietly

Double Negatives

A *double negative* is a negative statement that includes two negative elements. This is incorrect in Standard English.

> Incorrect: She hasn't never come to my house to visit.

> Correct: She has never come to my house to visit.

The intended meaning is that she has never come to the house, so the double negative is incorrect. However, it is possible to use two negatives to create a positive statement.

> Correct: She was not unhappy with her performance on the quiz.

In this case, the double negative, *was not unhappy*, is intended to show a positive, so it is correct. This means that she was somewhat happy with her performance.

Faulty Parallelism

It is necessary to use parallel construction in sentences that have multiple similar ideas. Using parallel structure provides clarity in writing. *Faulty parallelism* is created when multiple ideas are joined using different sentence structures. Compare these examples:

> Incorrect: We start each practice with stretches, a run, and fielding grounders.
> Correct: We start each practice with stretching, running, and fielding grounders.

> Incorrect: I watched some television, reading my book, and fell asleep.
> Correct: I watched some television, read my book, and fell asleep.

Incorrect: *Some of the readiness skills for Kindergarten are to cut with scissors, to tie shoes, and dressing independently.*

Correct: *Some of the readiness skills for Kindergarten are being able to cut with scissors, to tie shoes, and to dress independently.*

Subordination

If multiple pieces of information in a sentence are not equal, they can be joined by creating an independent clause and a dependent clause. The less important information becomes the *subordinate clause*:

Draft: The hotel was acceptable. We wouldn't stay at the hotel again.

Revised: Though the hotel was acceptable, we wouldn't stay there again.

The more important information (*we wouldn't stay there again*) becomes the main clause, and the less important information (*the hotel was acceptable*) becomes the subordinate clause.

Context Clues

Context clues help readers understand unfamiliar words, and thankfully, there are many types.

Synonyms are words or phrases that have nearly, if not exactly, the same meaning as other words or phrases

Large boxes are needed to pack *big* items.

Antonyms are words or phrases that have opposite definitions. Antonyms, like synonyms, can serve as context clues, although more cryptically.

Large boxes are not needed to pack *small* items.

Definitions are sometimes included within a sentence to define uncommon words.

They practiced the *rumba*, a *type of dance*, for hours on end.

Explanations provide context through elaboration.

Large boxes holding items weighing over 60 pounds were stacked in the corner.

Here's an example of *contrast*:

These *minute* creatures were much different than the *huge* mammals that the zoologist was accustomed to dealing with.

Beware of Simplicity

Sometimes the answer may seem very simple. In this case, it's prudent to look more carefully at the question and the possible answer choices. Very brief answers aren't always correct, and the opposite may also be true. The goal is to read all the answer choices carefully, trying to rule out those that don't make sense.

Final Notes

It's best to read every answer choice before making a decision. While some answers may seem plausible, there may be others that are better choices. First instinct is usually right, but reading every answer is recommended. Caution should be taken in choosing an answer that "sounds right." Grammar rules can be tricky, and what sounds right may not be correct. It's best to rely on knowledge of grammar to choose the best answer. Ruling out incorrect responses can help narrow the choices down. Choosing between two choices (after reading them carefully) and selecting the answer that best matches the rules of Standard English is less overwhelming.

Practice Questions

Read the selection about travelling in an RV and answer Questions 1-5.

(1) I have to admit that when my father bought a recreational vehicle (RV), I thought he was making a huge mistake. (2) I didn't really know anything about RVs, but I knew that my dad was as big a "city slicker" as there was. (3) In fact, I even thought he might have gone a little bit crazy. (4) On trips to the beach, he preferred to swim at the pool, and whenever he went hiking, he avoided touching any plants for fear that they might be poison ivy. (5) Why would this man, with an almost irrational fear of the outdoors, want a 40-foot camping behemoth?

(6) The RV was a great purchase for our family and brought us all closer together. (7) Every morning we would wake up, eat breakfast, and broke camp. (8) We laughed at our own comical attempts to back The Beast into spaces that seemed impossibly small. (9) We rejoiced as "hackers." (10) When things inevitably went wrong and we couldn't solve the problems on our own, we discovered the incredible helpfulness and friendliness of the RV community. (11) We even made some new friends in the process.

(12) Above all, it allowed us to share adventures. (13) While travelling across America, which we could not have experienced in cars and hotels. (14) Enjoying a campfire on a chilly summer evening with the mountains of Glacier National Park in the background, or waking up early in the morning to see the sun rising over the distant spires of Arches National Park are memories that will always stay with me and our entire family.

1. Which of the following would be the best choice for this sentence?
   a. Leave it where it is now.
   b. Move the sentence so that it comes before the preceding sentence.
   c. Move the sentence to the end of the first paragraph.
   d. Omit the sentence.

2. Which of the following would be the best choice for the underlined portion of Sentence 6 (reproduced below)?

   <u>The RV</u> was a great purchase for our family and brought us all closer together.

   a. NO CHANGE
   b. Not surprisingly, the RV
   c. Furthermore, the RV
   d. As it turns out, the RV

3. Which of the following would be the best choice for the underlined portion of Sentence 7 (reproduced below)?

Every morning <u>we would wake up, eat breakfast, and broke camp.</u>

a. NO CHANGE
b. we would wake up, eat breakfast, and break camp.
c. would we wake up, eat breakfast, and break camp?
d. we are waking up, eating breakfast, and breaking camp.

4. Which of the following would be the best choice for Sentence 9 (reproduced below)?

We rejoiced as "hackers."

a. NO CHANGE
b. To a nagging problem of technology, we rejoiced as "hackers."
c. We rejoiced when we figured out how to "hack" a solution to a nagging technological problem.
d. To "hack" our way to a solution, we had to rejoice.

5. Which of the following would be the best choice for Sentences 12 and 13 (reproduced below)?

<u>Above all, it allowed us to share adventures. While travelling across America,</u> which we could not have experienced in cars and hotels.

a. NO CHANGE
b. Above all, it allowed us to share adventures while traveling across America
c. Above all, it allowed us to share adventures; while traveling across America
d. Above all, it allowed us to share adventures—while traveling across America

Answer Explanations

1. B: Move the sentence so that it comes before the preceding sentence. For this question, place the underlined sentence in each prospective choice's position. To keep as-is is incorrect because the father "going crazy" doesn't logically follow the fact that he was a "city slicker." Choice C is incorrect because the sentence in question is not a concluding sentence and does not transition smoothly into the second paragraph. Choice *D* is incorrect because the sentence doesn't necessarily need to be omitted since it logically follows the very first sentence in the passage.

2. D: Choice *D* is correct because "As it turns out" indicates a contrast from the previous sentiment, that the RV was a great purchase. Choice *A* is incorrect because the sentence needs an effective transition from the paragraph before. Choice *B* is incorrect because the text indicates it *is* surprising that the RV was a great purchase because the author was skeptical beforehand. Choice *C* is incorrect because the transition "Furthermore" does not indicate a contrast.

3. B: This sentence calls for parallel structure. Choice *B* is correct because the verbs "wake," "eat," and "break" are consistent in tense and parts of speech. Choice *A* is incorrect because the words "wake" and "eat" are present tense while the word "broke" is in past tense. Choice *C* is incorrect because this turns the sentence into a question, which doesn't make sense within the context. Choice *D* is incorrect because it breaks tense with the rest of the passage. "Waking," "eating," and "breaking" are all present participles, and the context around the sentence is in past tense.

4. C: Choice *C* is correct because it is clear and fits within the context of the passage. Choice *A* is incorrect because "We rejoiced as 'hackers'" does not give a reason why hacking was rejoiced. Choice *B* is incorrect because it does not mention a solution being found and is therefore not specific enough. Choice *D* is incorrect because the sentence does not give enough detail as to what the problem entails.

5. B: Choice *B* is correct because there is no punctuation needed if a dependent clause ("while traveling across America") is located behind the independent clause ("it allowed us to share adventures"). Choice *A* is incorrect because there are two dependent clauses connected and no independent clause, and a complete sentence requires at least one independent clause. Choice *C* is incorrect because of the same reason as Choice *A*. Semicolons have the same function as periods: there must be an independent clause on either side of the semicolon. Choice *D* is incorrect because the dash simply interrupts the complete sentence.

WriterPlacer (Written Essay)

Brainstorming

One of the most important steps in writing an essay is prewriting. Before drafting an essay, it's helpful to think about the topic for a moment or two, in order to gain a more solid understanding of what the task is. Then, spending about five minutes jotting down the immediate ideas that could work for the essay is recommended. It is a way to get some words on the page and offer a reference for ideas when drafting. Scratch paper is provided for writers to use any prewriting techniques such as webbing, free writing, or listing. The goal is to get ideas out of the mind and onto the page.

Considering Opposing Viewpoints

In the planning stage, it's important to consider all aspects of the topic, including different viewpoints on the subject. There are more than two ways to look at a topic, and a strong argument considers those opposing viewpoints. Considering opposing viewpoints can help writers present a fair, balanced, and informed essay that shows consideration for all readers. This approach can also strengthen an argument by recognizing and potentially refuting the opposing viewpoint(s).

Drawing from personal experience may help to support ideas. For example, if the goal for writing is a personal narrative, then the story should be from the writer's own life. Many writers find it helpful to draw from personal experience, even in an essay that is not strictly narrative. Personal anecdotes or short stories can help to illustrate a point in other types of essays as well.

Moving from Brainstorming to Planning

Once the ideas are on the page, it's time to turn them into a solid plan for the essay. The best ideas from the brainstorming results can then be developed into a more formal outline. An outline typically has one main point (the thesis) and at least three sub-points that support the main point. Here's an example:

Main Idea

- Point #1
- Point #2
- Point #3

Of course, there will be details under each point, but this approach is the best for dealing with timed writing.

Staying on Track

Basing the essay on the outline aids in both organization and coherence. The goal is to ensure that there is enough time to develop each sub-point in the essay, roughly spending an equal amount of time on each idea. Keeping an eye on the time will help. If there are fifteen minutes left to draft the essay, then it makes sense to spend about 5 minutes on each of the ideas. Staying on task is critical to success, and timing out the parts of the essay can help writers avoid feeling overwhelmed.

Parts of the Essay

The *introduction* has to do a few important things:

- Establish the *topic* of the essay in original wording (i.e., not just repeating the prompt)
- Clarify the significance/importance of the topic or purpose for writing (not too many details, a brief overview)
- Offer a *thesis statement* that identifies the writer's own viewpoint on the topic (typically one-two brief sentences as a clear, concise explanation of the main point on the topic)

Body paragraphs reflect the ideas developed in the outline. Three-four points is probably sufficient for a short essay, and they should include the following:

- A *topic sentence* that identifies the sub-point (e.g., a reason why, a way how, a cause or effect)
- A detailed *explanation* of the point, explaining why the writer thinks this point is valid
- Illustrative examples, such as personal examples or real-world examples, that support and validate the point (i.e., "prove" the point)
- A *concluding sentence* that connects the examples, reasoning, and analysis to the point being made

The *conclusion*, or final paragraph, should be brief and should reiterate the focus, clarifying why the discussion is significant or important. It is important to avoid adding specific details or new ideas to this paragraph. The purpose of the conclusion is to sum up what has been said to bring the discussion to a close.

Don't Panic!

Writing an essay can be overwhelming, and performance panic is a natural response. The outline serves as a basis for the writing and helps to keep writers focused. Getting stuck can also happen, and it's helpful to remember that brainstorming can be done at any time during the writing process. Following the steps of the writing process is the best defense against writer's block.

Timed essays can be particularly stressful, but assessors are trained to recognize the necessary planning and thinking for these timed efforts. Using the plan above and sticking to it helps with time management. Timing each part of the process helps writers stay on track. Sometimes writers try to cover too much in their essays. If time seems to be running out, this is an opportunity to determine whether all of the ideas in the outline are necessary. Three body paragraphs are sufficient, and more than that is probably too much to cover in a short essay.

More isn't always *better* in writing. A strong essay will be clear and concise. It will avoid unnecessary or repetitive details. It is better to have a concise, five-paragraph essay that makes a clear point, than a ten-paragraph essay that doesn't. The goal is to write one-two pages of quality writing. Paragraphs should also reflect balance; if the introduction goes to the bottom of the first page, the writing may be going off-track or be repetitive. It's best to fall into the one-two page range, but a complete, well-developed essay is the ultimate goal.

The Final Steps

Leaving a few minutes at the end to revise and proofread offers an opportunity for writers to polish things up. Putting one's self in the reader's shoes and focusing on what the essay actually says helps writers identify problems—it's a movement from the mindset of writer to the mindset of editor. The goal is to have a clean, clear copy of the essay. The following areas should be considered when proofreading:

- Sentence fragments
- Awkward sentence structure
- Run-on sentences
- Incorrect word choice
- Grammatical agreement errors
- Spelling errors
- Punctuation errors
- Capitalization errors

The Short Overview

The essay may seem challenging, but following these steps can help writers focus:

- Take one-two minutes to think about the topic.
- Generate some ideas through brainstorming (three-four minutes).
- Organize ideas into a brief outline, selecting just three-four main points to cover in the essay (eventually the body paragraphs).
- Develop essay in parts:
- Introduction paragraph, with intro to topic and main points
- Viewpoint on the subject at the end of the introduction
- Body paragraphs, based on outline
- Each paragraph: makes a main point, explains the viewpoint, uses examples to support the point
- Brief conclusion highlighting the main points and closing
- Read over the essay (last five minutes).
- Look for any obvious errors, making sure that the writing makes sense.

Practice Prompt

Prepare an essay of about 300-600 words on the topic below.

Some people feel that sharing their lives on social media sites such as Facebook, Instagram, and Snapchat is fine. They share every aspect of their lives, including pictures of themselves and their families, what they ate for lunch, who they are dating, and when they are going on vacation. They even say that if it's not on social media, it didn't happen. Other people believe that sharing so much personal information is an invasion of privacy and could prove dangerous. They think sharing personal pictures and details invites predators, cyberbullying, and identity theft.

Write an essay to someone who is considering whether to participate in social media. Take a side on the issue and argue whether or not he/she should join a social media network. Use specific examples to support your argument.

Practice Test #1

Arithmetic

1. 3.4+2.35+4=
    a. 5.35
    b. 9.2
    c. 9.75
    d. 10.25

2. $5.88 \times 3.2 =$
    a. 18.816
    b. 16.44
    c. 20.352
    d. 17

3. $\frac{3}{25} =$
    a. 0.15
    b. 0.1
    c. 0.9
    d. 0.12

4. Which of the following is largest?
    a. 0.45
    b. 0.096
    c. 0.3
    d. 0.313

5. Which of the following is NOT a way to write 40 percent of N?
    a. $(0.4)N$
    b. $\frac{2}{5}N$
    c. $40N$
    d. $\frac{4N}{10}$

6. Which is closest to 17.8×9.9?
    a. 140
    b. 180
    c. 200
    d. 350

7. A student gets an 85% on a test with 20 questions. How many answers did the student solve correctly?
    a. 15
    b. 16
    c. 17
    d. 18

8. Four people split a bill. The first person pays for $\frac{1}{5}$, the second person pays for $\frac{1}{4}$, and the third person pays for $\frac{1}{3}$. What fraction of the bill does the fourth person pay?

   a. $\frac{13}{60}$

   b. $\frac{47}{60}$

   c. $\frac{1}{4}$

   d. $\frac{4}{15}$

9. 6 is 30% of what number?

   a. 18
   b. 20
   c. 24
   d. 26

10. $3\frac{2}{3} - 1\frac{4}{5} =$

   a. $1\frac{13}{15}$

   b. $\frac{14}{15}$

   c. $2\frac{2}{3}$

   d. $\frac{4}{5}$

11. What is $\frac{420}{98}$ rounded to the nearest integer?

   a. 4
   b. 3
   c. 5
   d. 6

12. $4\frac{1}{3} + 3\frac{3}{4} =$

   a. $6\frac{5}{12}$

   b. $8\frac{1}{12}$

   c. $8\frac{2}{3}$

   d. $7\frac{7}{12}$

13. Five of six numbers have a sum of 25. The average of all six numbers is 6. What is the sixth number?

   a. 8
   b. 10
   c. 11
   d. 12

14. $52.3 \times 10^{-3} =$
    a. 0.00523
    b. 0.0523
    c. 0.523
    d. 523

15. If $\frac{5}{2} \div \frac{1}{3} = n$, then n is between:
    a. 5 and 7
    b. 7 and 9
    c. 9 and 11
    d. 3 and 5

16. A closet is filled with red, blue, and green shirts. If $\frac{1}{3}$ of the shirts are green and $\frac{2}{5}$ are red, what fraction of the shirts are blue?
    a. $\frac{4}{15}$

    b. $\frac{1}{5}$

    c. $\frac{7}{15}$

    d. $\frac{1}{2}$

17. Shawna buys $2\frac{1}{2}$ gallons of paint. If she uses $\frac{1}{3}$ of it on the first day, how much does she have left?
    a. $1\frac{5}{6}$ gallons

    b. $1\frac{1}{2}$ gallons

    c. $1\frac{2}{3}$ gallons

    d. 2 gallons

18. Mom's car drove 72 miles in 90 minutes. How fast did she drive in feet per second?
    a. 0.8 feet per second
    b. 48.9 feet per second
    c. 0.009 feet per second
    d. 70.4 feet per second

19. What is the simplified form of the expression $1.2 \times 10^{12} \div 3.0 \times 10^8$?
    a. 0.4×10^4
    b. 4.0×10^4
    c. 4.0×10^3
    d. 3.6×10^{20}

20. You measure the width of your door to be 36 inches. The true width of the door is 35.75 inches. What is the relative error in your measurement?
    a. 0.7%
    b. 0.007%
    c. 0.99%
    d. 0.1%

Quantitative Reasoning, Algebra, and Statistics

1. Which of the following is the result of simplifying the expression: $\frac{4a^{-1}b^3}{a^4b^{-2}} \times \frac{3a}{b}$?
    a. $12a^3b^5$
    b. $12\frac{b^4}{a^4}$
    c. $\frac{12}{a^4}$
    d. $7\frac{b^4}{a}$

2. Write the expression for three times the sum of twice a number and one minus 6.
    a. $2x + 1 - 6$
    b. $3x + 1 - 6$
    c. $3(x + 1) - 6$
    d. $3(2x + 1) - 6$

3. On Monday, Robert mopped the floor in 4 hours. On Tuesday, he did it in 3 hours. If on Monday, his average rate of mopping was p sq. ft. per hour, what was his average rate on Tuesday?
    a. $\frac{4}{3}p$ sq. ft. per hour

    b. $\frac{3}{4}p$ sq. ft. per hour

    c. $\frac{5}{4}p$ sq. ft. per hour

    d. $p + 1$ sq. ft. per hour

4. The phone bill is calculated each month using the equation $c = 50g + 75$. The cost of the phone bill per month is represented by c, and g represents the gigabytes of data used that month. What is the value and interpretation of the slope of this equation?

    a. 75 dollars per day

    b. 75 gigabytes per day

    c. 50 dollars per day

    d. 50 dollars per gigabyte

5. For the following similar triangles, what are the values of x and y (rounded to one decimal place)?

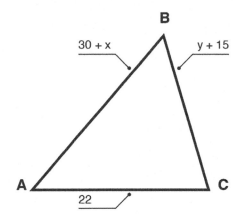
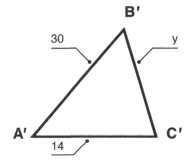

    a. $x = 16.5, y = 25.1$

    b. $x = 19.5, y = 24.1$

    c. $x = 17.1, y = 26.3$

    d. $x = 26.3, y = 17.1$

6. Solve for x, if $x^2 - 2x - 8 = 0$.

    a. $2 \pm \frac{\sqrt{30}}{2}$

    b. $2 \pm 4\sqrt{2}$

    c. 1 ± 3

    d. $4 \pm \sqrt{2}$

7. Which of the following is a factor of both $x^2 + 4x + 4$ and $x^2 - x - 6$?

    a. $x - 3$

    b. $x + 2$

    c. $x - 2$

    d. $x + 3$

8. What is the solution to the following system of equations?

$$x^2 - 2x + y = 8$$

$$x - y = -2$$

  a. $(-2, 3)$
  b. There is no solution.
  c. $(-2, 0) \ (1, 3)$
  d. $(-2, 0) \ (3, 5)$

9. What are the center and radius of a circle with equation $4x^2 + 4y^2 - 16x - 24y + 51 = 0$?
  a. Center $(3, 2)$ and radius $1/2$
  b. Center $(2, 3)$ and radius $1/2$
  c. Center $(3, 2)$ and radius $1/4$
  d. Center $(2, 3)$ and radius $1/4$

10. What's the midpoint of a line segment with endpoints $(-1, 2)$ and $(3, -6)$?
  a. $(1, 2)$
  b. $(1, 0)$
  c. $(-1, 2)$
  d. $(1, -2)$

11. A sample data set contains the following values: 1, 3, 5, 7. What's the standard deviation of the set?
  a. 2.58
  b. 4
  c. 6.23
  d. 1.1

12. A ball is drawn at random from a ball pit containing 8 red balls, 7 yellow balls, 6 green balls, and 5 purple balls. What's the probability that the ball drawn is yellow?

    a. $\frac{1}{26}$

    b. $\frac{19}{26}$

    c. $\frac{7}{26}$

    d. 1

13. Two cards are drawn from a shuffled deck of 52 cards. What's the probability that both cards are Kings if the first card isn't replaced after it's drawn?

    a. $\frac{1}{169}$

    b. $\frac{1}{221}$

    c. $\frac{1}{13}$

    d. $\frac{4}{13}$

14. What's the probability of rolling a 6 at least once in two rolls of a die?

    a. $\frac{1}{3}$

    b. $\frac{1}{36}$

    c. $\frac{1}{6}$

    d. $\frac{11}{36}$

15. A line goes through the point (-4, 0) and the point (0,2). What is the slope of the line?

    a. 2

    b. 4

    c. $\frac{3}{2}$

    d. $\frac{1}{2}$

16. Six people apply to work for Janice's company, but she only needs four workers. How many different groups of four employees can Janice choose?

    a. 6
    b. 10
    c. 15
    d. 36

17. $2x(3x + 1) - 5(3x + 1) =$

    a. $10x(3x + 1)$

    b. $10x^2(3x + 1)$

    c. $(2x - 5)(3x + 1)$

    d. $(2x + 1)(3x - 5)$

18. For which real numbers x is $-3x^2 + x - 8 > 0$?
    a. All real numbers x

    b. $-2\sqrt{\frac{2}{3}} < x < 2\sqrt{\frac{2}{3}}$

    c. $1 - 2\sqrt{\frac{2}{3}} < x < 1 + 2\sqrt{\frac{2}{3}}$

    d. For no real numbers x

19. A root of $x^2 - 2x - 2$ is
    a. $1 + \sqrt{3}$
    b. $1 + 2\sqrt{2}$
    c. $2 + 2\sqrt{3}$
    d. $2 - 2\sqrt{3}$

20. In the xy-plane, the graph of $y = x^2 + 2$ and the circle with center $(0,1)$ and radius 1 have how many points of intersection?
    a. 0
    b. 1
    c. 2
    d. 3

Advanced Algebra and Functions

1. What is the value of $x^2 - 2xy + 2y^2$ when $x = 2, y = 3$?
    a. 8
    b. 10
    c. 12
    d. 14

2. $(2x - 4y)^2 =$
    a. $4x^2 - 16xy + 16y^2$
    b. $4x^2 - 8xy + 16y^2$
    c. $4x^2 - 16xy - 16y^2$
    d. $2x^2 - 8xy + 8y^2$

3. If $x > 3$, then $\frac{x^2 - 6x + 9}{x^2 - x - 6} =$

    a. $\frac{x+2}{x-3}$

    b. $\frac{x-2}{x-3}$

    c. $\frac{x-3}{x+3}$

    d. $\frac{x-3}{x+2}$

4. The square and circle have the same center. The circle has a radius of r. What is the area of the shaded region?

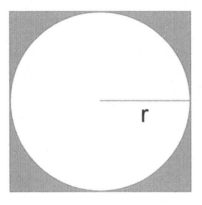

a. $r^2 - \pi r^2$
b. $4r^2 - 2\pi r$
c. $(4 - \pi)r^2$
d. $(\pi - 1)r^2$

5. If $4x - 3 = 5$, then $x =$
a. 1
b. 2
c. 3
d. 4

6. What are the zeros of the function: $f(x) = x^3 + 4x^2 + 4x$?
a. −2
b. 0, −2
c. 2
d. 0, 2

7. If $g(x) = x^3 - 3x^2 - 2x + 6$ and $f(x) = 2$, then what is $g(f(x))$?
a. −26
b. 6
c. $2x^3 - 6x^2 - 4x + 12$
d. −2

8. What is the function that forms an equivalent graph to $y = \cos(x)$?
a. $y = \tan(x)$
b. $y = \csc(x)$
c. $y = \sin(x + \frac{\pi}{2})$
d. $y = \sin(x - \frac{\pi}{2})$

9. What is the solution for the equation $\tan(x) + 1 = 0$, where $0 \leq x < 2\pi$?

    a. $x = \dfrac{3\pi}{4}, \dfrac{5\pi}{4}$

    b. $x = \dfrac{3\pi}{4}, \dfrac{\pi}{4}$

    c. $x = \dfrac{5\pi}{4}, \dfrac{7\pi}{4}$

    d. $x = \dfrac{3\pi}{4}, \dfrac{7\pi}{4}$

10. Which of the following inequalities is equivalent to $3 - \dfrac{1}{2}x \geq 2$?

    a. $x \geq 2$
    b. $x \leq 2$
    c. $x \geq 1$
    d. $x \leq 1$

11. For which of the following are $x = 4$ and $x = -4$ solutions?

    a. $x^2 + 16 = 0$
    b. $x^2 + 4x - 4 = 0$
    c. $x^2 - 2x - 2 = 0$
    d. $x^2 - 16 = 0$

12. If x is not zero, then $\dfrac{3}{x} + \dfrac{5u}{2x} - \dfrac{u}{4} =$

    a. $\dfrac{12+10u-ux}{4x}$

    b. $\dfrac{3+5u-ux}{x}$

    c. $\dfrac{12x+10u+ux}{4x}$

    d. $\dfrac{12+10u-u}{4x}$

13. Which of the following shows the correct result of simplifying the following expression?

$$(7n + 3n^3 + 3) + (8n + 5n^3 + 2n^4)$$

    a. $9n^4 + 15n - 2$
    b. $2n^4 + 5n^3 + 15n - 2$
    c. $9n^4 + 8n^3 + 15n$
    d. $2n^4 + 8n^3 + 15n + 3$

14. How could the following equation be factored to find the zeros?

$$y = x^3 - 3x^2 - 4x$$

    a. $0 = x^2(x - 4), x = 0, 4$
    b. $0 = 3x(x + 1)(x + 4), x = 0, -1, -4$
    c. $0 = x(x + 1)(x + 6), x = 0, -1, -6$
    d. $0 = x(x + 1)(x - 4), x = 0, -1, 4$

15. What is the simplified quotient of $\frac{5x^3}{3x^2y} \div \frac{25}{3y^9}$?

    a. $\frac{125x}{9y^{10}}$

    b. $\frac{x}{5y^8}$

    c. $\frac{5}{xy^8}$

    d. $\frac{xy^8}{5}$

16. What type of function is modeled by the values in the following table?

x	$f(x)$
1	2
2	4
3	8
4	16
5	32

    a. Linear
    b. Exponential
    c. Quadratic
    d. Cubic

17. What is the simplified form of the expression $tan\theta \ cos\theta$?
    a. $sin\theta$
    b. 1
    c. $csc\theta$
    d. $\dfrac{1}{sec\theta}$

18. Which equation is not a function?
    a. $y = |x|$
    b. $y = \sqrt{x}$
    c. $x = 3$
    d. $y = 4$

19. How could the following function be rewritten to identify the zeros?

$$y = 3x^3 + 3x^2 - 18x$$

    a. $y = 3x(x + 3)(x - 2)$
    b. $y = x(x - 2)(x + 3)$
    c. $y = 3x(x - 3)(x + 2)$
    d. $y = (x + 3)(x - 2)$

20. If $x^2 + x - 3 = 0$, then $\left(x - \frac{1}{2}\right)^2 =$
    a. $\dfrac{11}{2}$

    b. $\dfrac{11}{4}$

    c. 11

    d. $\dfrac{121}{4}$

Reading Comprehension

Questions 1–6 are based upon the following passage:

This excerpt is adaptation from "What to the Slave is the Fourth of July?" Rochester, New York July 5, 1852

Fellow citizens—Pardon me, and allow me to ask, why am I called upon to speak here today? What have I, or those I represent, to do with your national independence? Are the great principles of political freedom and of natural justice, embodied in that Declaration of Independence, extended to us? And am I therefore called upon to bring our humble offering to the national altar, and to confess the benefits, and express devout gratitude for the blessings, resulting from your independence to us?

Would to God, both for your sakes and ours, ours that an affirmative answer could be truthfully returned to these questions! Then would my task be light, and my burden easy and delightful. For who is there so cold that a nation's sympathy could not warm him? Who so obdurate and dead to the claims of gratitude that would not thankfully acknowledge such priceless benefits? Who so stolid and selfish, that would not give his voice to swell the hallelujahs of a nation's jubilee, when the chains of servitude had been torn from his limbs? I am not that man. In a case like that, the dumb may eloquently speak, and the lame man leap as a hart.

But, such is not the state of the case. I say it with a sad sense of the disparity between us. I am not included within the pale of this glorious anniversary. Oh pity! Your high independence only reveals the immeasurable distance between us. The blessings in which you this day rejoice, I do not enjoy in common. The rich inheritance of justice, liberty, prosperity, and independence, bequeathed by your fathers, is shared by *you*, not by *me*. This Fourth of July is *yours*, not *mine*. You may rejoice, *I* must mourn. To drag a man in fetters into the grand illuminated temple of liberty, and call upon him to join you in joyous anthems, were inhuman mockery and sacrilegious irony. Do you mean, citizens, to mock me, by asking me to speak today? If so there is a parallel to your conduct. And let me warn you that it is dangerous to copy the example of a nation whose crimes, towering up to heaven, were thrown down by the breath of the Almighty, burying that nation and irrecoverable ruin! I can today take up the plaintive lament of a peeled and woe-smitten people.

By the rivers of Babylon, there we sat down. Yea! We wept when we remembered Zion. We hanged our harps upon the willows in the midst thereof. For there, they that carried us away captive, required of us a song; and they who wasted us required of us mirth, saying, "Sing us one of the songs of Zion." How can we sing the Lord's song in a strange land? If I forget thee, O Jerusalem, let my right hand forget her cunning. If I do not remember thee, let my tongue cleave to the roof of my mouth.

1. What is the tone of the first paragraph of this passage?
   a. Exasperated
   b. Inclusive
   c. Contemplative
   d. Nonchalant

2. Which word CANNOT be used synonymously with the term *obdurate* as it is conveyed in the text below?

> Who so obdurate and dead to the claims of gratitude, that would not thankfully acknowledge such priceless benefits?

a. Steadfast
b. Stubborn
c. Contented
d. Unwavering

3. What is the central purpose of this text?
a. To demonstrate the author's extensive knowledge of the Bible
b. To address the hypocrisy of the Fourth of July holiday
c. To convince wealthy landowners to adopt new holiday rituals
d. To explain why minorities often relished the notion of segregation in government institutions

4. Which statement serves as evidence of the previous question?
a. By the rivers of Babylon . . . down.
b. Fellow citizens . . . today.
c. I can . . . woe-smitten people.
d. The rich inheritance of justice . . . *not by me*.

5. The statement below features an example of which of the following literary devices?

> Oh pity! Your high independence only reveals the immeasurable distance between us.

a. Assonance
b. Parallelism
c. Amplification
d. Hyperbole

6. The speaker's use of biblical references, such as "rivers of Babylon" and the "songs of Zion," helps the reader to do all of the following EXCEPT which of the following?
a. Identify with the speaker through the use of common text.
b. Convince the audience that injustices have been committed by referencing another group of people who have been previously affected by slavery.
c. Display the equivocation of the speaker and those that he represents.
d. Appeal to the listener's sense of humanity.

7. Which organizational style is used in the following passage?

> There are several reasons why the new student café has not been as successful as expected. One factor is that prices are higher than originally advertised, so many students cannot afford to buy food and beverages there. Also, the café closes rather early; as a result, students go out in town to other late-night gathering places rather than meeting friends at the café on campus.

a. Cause and effect order
b. Compare and contrast order
c. Spatial order
d. Time order

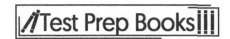

8. What can the reader infer from this passage?
   I would sometimes lie down, and let five or six of them dance on my hand; and at last the boys and girls would venture to come and play at hide-and-seek in my hair.

   a. The children tortured Gulliver.
   b. Gulliver traveled because he wanted to meet new people.
   c. Gulliver is considerably larger than the children who are playing around him.
   d. Gulliver has a genuine love and enthusiasm for people of all sizes.

9. David Foster Wallace's *Infinite Jest* is the holy grail of modern literature. It will stand the test of time in its relevance. Every single person who starts reading *Infinite Jest* cannot physically put down the book until completing it.

Which of the following is the main point of the passage?
   a. David Foster Wallace's *Infinite Jest* is the holy grail of modern literature.
   b. *Infinite Jest* is a page-turner.
   c. David Foster Wallace wrote *Infinite Jest*.
   d. *Infinite Jest* is a modern classic for good reason, and everybody should read it.

10. The assassination of Archduke Franz Ferdinand of Austria is often ascribed as the cause of World War I. However, the assassination merely lit the fuse in a combustible situation since many of the world powers were in complicated and convoluted military alliances. For example, England, France, and Russia entered into a mutual defense treaty seven years prior to World War I. Even without Franz Ferdinand's assassination _____.

Which of the following most logically completes the passage?
   a. A war between the world powers was extremely likely.
   b. World War I never would have happened.
   c. England, France, and Russia would have started the war.
   d. Austria would have started the war.

Questions 11–15 are based on the following passages.

Passage I

Lethal force, or deadly force, is defined as the physical means to cause death or serious harm to another individual. The law holds that lethal force is only accepted when you or another person are in immediate and unavoidable danger of death or severe bodily harm. For example, a person could be beating a weaker person in such a way that they are suffering severe enough trauma that could result in death or serious harm. This would be an instance where lethal force would be acceptable and possibly the only way to save that person from irrevocable damage.

Another example of when to use lethal force would be when someone enters your home with a deadly weapon. The intruder's presence and possession of the weapon indicate mal-intent and the ability to inflict death or severe injury to you and your loved ones. Again, lethal force can be used in this situation. Lethal force can also be applied to prevent the harm of another individual. If a woman is being brutally assaulted and is unable to fend off an attacker, lethal force can be used to defend her as a last-ditch effort. If she is in immediate jeopardy of rape, harm, and/or death, lethal force could be the only response that could effectively deter the assailant.

259

The key to understanding the concept of lethal force is the term *last resort*. Deadly force cannot be taken back; it should be used only to prevent severe harm or death. The law does distinguish whether the means of one's self-defense is fully warranted, or if the individual goes out of control in the process. If you continually attack the assailant after they are rendered incapacitated, this would be causing unnecessary harm, and the law can bring charges against you. Likewise, if you kill an attacker unnecessarily after defending yourself, you can be charged with murder. This would move lethal force beyond necessary defense, making it no longer a last resort but rather a use of excessive force.

Passage II

Assault is the unlawful attempt of one person to apply apprehension on another individual by an imminent threat or by initiating offensive contact. Assaults can vary, encompassing physical strikes, threatening body language, and even provocative language. In the case of the latter, even if a hand has not been laid, it is still considered an assault because of its threatening nature.

Let's look at an example: A homeowner is angered because his neighbor blows fallen leaves into his freshly mowed lawn. Irate, the homeowner gestures a fist to his fellow neighbor and threatens to bash his head in for littering on his lawn. The homeowner's physical motions and verbal threat heralds a physical threat against the other neighbor. These factors classify the homeowner's reaction as an assault. If the angry neighbor hits the threatening homeowner in retaliation, that would constitute an assault as well because he physically hit the homeowner.

Assault also centers on the involvement of weapons in a conflict. If someone fires a gun at another person, it could be interpreted as an assault unless the shooter acted in self-defense. If an individual drew a gun or a knife on someone with the intent to harm them, it would be considered assault. However, it's also considered an assault if someone simply aimed a weapon, loaded or not, at another person in a threatening manner.

11. What is the purpose of the second passage?
    a. To inform the reader about what assault is and how it is committed.
    b. To inform the reader about how assault is a minor example of lethal force.
    c. To argue that the use of assault is more common than the use of lethal force.
    d. The author is recounting an incident in which they were assaulted.

12. Which of the following situations, according to the passages, would not constitute an illegal use of lethal force?
    a. A disgruntled cashier yells obscenities at a customer.
    b. A thief is seen running away with stolen cash.
    c. A man is attacked in an alley by another man with a knife.
    d. A woman punches another woman in a bar.

13. Given the information in the passages, which of the following must be true about assault?
    a. All assault is considered expression of lethal force.
    b. There are various forms of assault.
    c. Assault is justified only as a last resort.
    d. Assault charges are more severe than unnecessary use of force charges.

14. Which of the following, if true, would most seriously undermine the explanation proposed by the author in Passage I, third paragraph?

    a. An instance of lethal force in self-defense is not absolutely absolved from blame. The law takes into account the necessary use of force at the time it is committed.

    b. An individual who uses necessary defense under lethal force is in direct compliance of the law under most circumstances.

    c. Lethal force in self-defense should be forgiven in all cases for the peace of mind of the primary victim.

    d. The use of lethal force is not evaluated on the intent of the user but rather the severity of the primary attack that warranted self-defense.

15. Based on the passages, what can we infer about the relationship between assault and lethal force?

    a. An act of lethal force always leads to a type of assault.

    b. Assault and lethal force have no conceivable connection.

    c. An assault with deadly intent can lead to an individual using lethal force to preserve their well-being.

    d. If someone uses self-defense in a conflict, it is called deadly force; if actions or threats are intended, it is called assault.

Questions 16–20 are based upon the following passage:

This excerpt is adapted from *Our Vanishing Wildlife,* by William T. Hornaday

    Three years ago, I think there were not many bird-lovers in the United States, who believed it possible to prevent the total extinction of both egrets from our fauna. All the known rookeries accessible to plume-hunters had been totally destroyed. Two years ago, the secret discovery of several small, hidden colonies prompted William Dutcher, President of the National Association of Audubon Societies, and Mr. T. Gilbert Pearson, Secretary, to attempt the protection of those colonies. With a fund contributed for the purpose, wardens were hired and duly commissioned. As previously stated, one of those wardens was shot dead in cold blood by a plume hunter. The task of guarding swamp rookeries from the attacks of money-hungry desperadoes to whom the accursed plumes were worth their weight in gold, is a very chancy proceeding. There is now one warden in Florida who says that "before they get my rookery they will first have to get me."

    Thus far the protective work of the Audubon Association has been successful. Now there are twenty colonies, which contain all told, about 5,000 egrets and about 120,000 herons and ibises which are guarded by the Audubon wardens. One of the most important is on Bird Island, a mile out in Orange Lake, central Florida, and it is ably defended by Oscar E. Baynard. To-day, the plume hunters who do not dare to raid the guarded rookeries are trying to study out the lines of flight of the birds, to and from their feeding-grounds, and shoot them in transit. Their motto is—"Anything to beat the law, and get the plumes." It is there that the state of Florida should take part in the war.

    The success of this campaign is attested by the fact that last year a number of egrets were seen in eastern Massachusetts—for the first time in many years. And so to-day the question is, can the wardens continue to hold the plume-hunters at bay?

16. The author's use of first person pronoun in the following text does NOT have which of the following effects?

> Three years ago, I think there were not many bird-lovers in the United States, who believed it possible to prevent the total extinction of both egrets from our fauna.

a. The phrase *I think* acts as a sort of hedging, where the author's tone is less direct and/or absolute.
b. It allows the reader to more easily connect with the author.
c. It encourages the reader to empathize with the egrets.
d. It distances the reader from the text by overemphasizing the story.

17. What purpose does the quote serve at the end of the first paragraph?
a. The quote shows proof of a hunter threatening one of the wardens.
b. The quote lightens the mood by illustrating the colloquial language of the region.
c. The quote provides an example of a warden protecting one of the colonies.
d. The quote provides much needed comic relief in the form of a joke.

18. What is the meaning of the word *rookeries* in the following text?

> To-day, the plume hunters who do not dare to raid the guarded rookeries are trying to study out the lines of flight of the birds, to and from their feeding-grounds, and shoot them in transit.

a. Houses in a slum area
b. A place where hunters gather to trade tools
c. A place where wardens go to trade stories
d. A colony of breeding birds

19. What is on Bird Island?
a. Hunters selling plumes
b. An important bird colony
c. Bird Island Battle between the hunters and the wardens
d. An important egret with unique plumes

20. What is the main purpose of the passage?
a. To persuade the audience to act in preservation of the bird colonies
b. To show the effect hunting egrets has had on the environment
c. To argue that the preservation of bird colonies has had a negative impact on the environment
d. To demonstrate the success of the protective work of the Audubon Association

Writing

Questions 1–7 are based on the following passage:

(1) One of the icon's of romantic and science fiction literature remains Mary Shelley's classic, *Frankenstein, or The Modern Prometheus*. (2) Schools throughout the world still teach the book in literature and philosophy courses. (3) Scientific communities also engage in discussion on the novel. (4) But why? (5) Besides the novel's engaging writing style the story's central theme remains highly relevant in a world of constant discovery and moral dilemmas. (6) Central to the core narrative is the struggle between enlightenment and the cost of overusing power.

(7) The subtitle, *The Modern Prometheus*, encapsulates the inner theme of the story more than the main title of *Frankenstein*. (8) As with many romantic writers, Shelley invokes the classical

myths and symbolism of Ancient Greece and Rome to high light core ideas. (9) Looking deeper into the myth of Prometheus sheds light not only on the character of Frankenstein but also poses a psychological dilemma to the audience. (10) Prometheus is the titan who gave fire to mankind. (11) However, more than just fire he gave people knowledge and power. (12) The power of fire advanced civilization. (13) Yet, for giving fire to man, Prometheus is punished by the gods bound to a rock and tormented for his act. (14) This is clearly a parallel to Frankenstein—he is the modern Prometheus.

(15) Frankenstein's quest for knowledge becomes an obsession. (16) It leads him to literally create new life, breaking the bounds of conceivable science to illustrate that man can create life out of nothing. (17) Yet he ultimately faltered as a creator, abandoning his progeny in horror of what he created. (18) Frankenstein then suffers his creature's wrath, the result of his pride, obsession for power and lack of responsibility.

(19) Shelley isn't condemning scientific achievement. (20) Rather, her writing reflects that science and discovery are good things, but, like all power, it must be used wisely. (21) The text alludes to the message that one must have reverence for nature and be mindful of the potential consequences. (22) Frankenstein did not take responsibility or even consider how his actions would affect others. (23) His scientific brilliance ultimately led to suffering.

1. Which of the following would be the best choice for the underlined portion of Sentence 1 (reproduced below)?

> <u>One of the icon's of romantic and science fiction literature</u> remains Mary Shelley's classic, Frankenstein, or The Modern Prometheus.

    a. Leave it as is
    b. One of the icons of romantic and science fiction literature
    c. One of the icon's of romantic, and science fiction literature,
    d. The icon of romantic and science fiction literature

2. Which of the following would be the best choice for the underlined portion of Sentence 5 (reproduced below)?

> Besides the novel's engaging <u>writing style the story's central theme</u> remains highly relevant in a world of constant discovery and moral dilemmas.

    a. Leave it as is
    b. writing style the central theme of the story
    c. writing style, the story's central theme
    d. the story's central theme's writing style

3. Which of the following would be the best choice for the underlined portion of Sentence 6 (reproduced below)?

Central to the core narrative is the <u>struggle between enlightenment and the cost of overusing power.</u>

a. Leave it as is
b. struggle between enlighten and the cost of overusing power.
c. struggle between enlightenment's cost of overusing power.
d. struggle between enlightening and the cost of overusing power.

4. Which of the following would be the best choice for the underlined portion of Sentence 8 (reproduced below)?

As with many romantic writers, Shelley invokes the classical myths and <u>symbolism of Ancient Greece and Rome to high light core ideas.</u>

a. Leave it as is
b. symbolism of Ancient Greece and Rome to highlight core ideas.
c. symbolism of ancient Greece and Rome to highlight core ideas.
d. symbolism of Ancient Greece and Rome highlighting core ideas.

5. Which of the following would be the best choice for the underlined portion of Sentence 9 (reproduced below)?

Looking deeper into the myth of Prometheus sheds light not only on the character of Frankenstein <u>but also poses a psychological dilemma to the audience.</u>

a. Leave it as is
b. but also poses a psychological dilemma with the audience.
c. but also poses a psychological dilemma for the audience.
d. but also poses a psychological dilemma there before the audience.

6. Which of the following would be the best choice for Sentence 11 (reproduced below)?

However, more than just fire he gave people knowledge and power.

a. Leave it as is
b. However, more than just fire he gave people, knowledge, and power.
c. However, more than just fire, he gave people knowledge and power.
d. Besides actual fire, Prometheus gave people knowledge and power.

7. Which of the following would be the best choice for the underlined portion of Sentence 17 (reproduced below)?

Yet, for giving fire to man, Prometheus is <u>punished by the gods bound to a rock and tormented for his act.</u>

a. Leave it as is
b. punished by the gods, bound to a rock and tormented for his act.
c. bound to a rock and tormented as punishment by the gods.
d. punished for his act by being bound to a rock and tormented as punishment from the gods.

Questions 8-11 are based on the following passage:

(1) The country of Japan is a chain of islands located along the coast of Asia. (2) Japan is made up of four major islands, with Honshu being the largest. (3) Honshu is also the seventh largest island in the world. (4) Underneath Japan, there are many volcanoes, some active and some dormant, but only about 108 of them are active today.

(5) There are many exciting places to visit in Japan. (6) Osaka is home to Dontobori, the main entertainment district. (7) The popular sport of sumo wrestling is often hosted in the town's arena. (8) People from all over Japan ride the Shinkansen or bullet train to Osaka to watch the events. (9) Hakone is famous for black eggs. (10) The town of Hakone sits on top of a mountain near Mount Fuji, the tallest mountain in Japan. (11) The eggs are boiled in a hot spring on the mountainside. (12) The natural water turns the shell of the egg black; however, it does not change the taste of it. (13) Tokyo is the capital city of Japan. (14) The town of Kyoto is full of shrines and temples, some lined with gold. (15) It is also a well-known place for geisha sightings.

(16) Traditional Japanese food consists mostly of fish and rice. (17) Sushi is raw fish and vegetables rolled in rice and wrapped in seaweed. (18) It is typically dipped in a sauce made of soy sauce and wasabi paste. (19) Tempura is a light, crispy batter used to fry fish, chicken, or vegetables. (20) Soba is another traditional Japanese dish. (21) It is a type of noodle that is oftentimes eaten cold and served in a broth made with soy sauce.

8. Which transition sentence should be added to the beginning of Sentence 16?
a. Typical Japanese lunch boxes are called *bentos* and usually consist of foods such as dried fruit, rice, and sushi.
b. Japanese cuisine is much different from American cuisine.
c. Yakisoba is a traditional Japanese noodle dish served with slices of beef.
d. Gyoza is a traditional Japanese dish that many Americans refer to as *pot stickers*.

9. Which sentence should be moved to follow Sentence 8 to correct the flow of information in the paragraph?
a. Sentence 5
b. Sentence 7
c. Sentence 10
d. Sentence 15

10. Which revision of Sentence 4 uses the most precise language?
    a. Japan sits just above 108 active volcanos.
    b. Japan has many volcanos, with only 108 being active.
    c. Active volcanos are common in Japan, especially the biggest 108.
    d. Japan lies on a chain of volcanos, and more than 108 of them are active.

11. Which sentence should be revised to correct an error in sentence structure?
    a. Sentence 4
    b. Sentence 8
    c. Sentence 12
    d. Sentence 14

Questions 12-14 are based on the following passage:

(1) Some people confuse Memorial Day with Veterans Day; however, there is a big difference. (2) Veterans Day is a day to celebrate all service men and women who have worked or still work to ensure the safety of our nation. (3) Memorial Day is a day of celebration and remembrance of those who lost their lives fighting for our country.

(4) Memorial Day, originally known as Decoration Day, first began on May 5, 1868, to celebrate the lives of those lost during the Civil War. (5) However, after World War I, the meaning of Decoration Day was changed to honor not just those who had fallen in the Civil War but all those who had fallen fighting for our country. (6) On Decoration Day, families of fallen soldiers visited their loved ones' gravesites to decorate them with flowers, such as the red poppy, which is the official flower of Memorial Day.

(7) In 1950, Decoration Day was renamed Memorial Day and declared a federal holiday by President Nixon in 1971. (8) Memorial Day is now celebrated on the last Monday in May. (9) All federal buildings, schools, post offices, and banks are closed.

12. Which transition phrase should be added to the beginning of sentence 9?
    a. On this day,
    b. So,
    c. Because
    d. If

13. Which sentence should be moved to follow sentence 1 to correct the flow of information in the paragraph?
    a. Sentence 3
    b. Sentence 4
    c. Sentence 7
    d. Sentence 8

14. Which revision of sentence 6 uses the most precise language?
    a. On Decoration Day, families visited graves to decorate them with red poppies.
    b. Red poppies are used to decorate the graves of fallen soldiers during Decoration Day celebrations.
    c. Fallen soldiers get their graves decorated with red poppies by their loved ones on Decoration Day.
    d. On Decoration Day, families of fallen soldiers visited graves and decorated them with flowers, such as the red poppy, the official flower of Memorial Day.

Questions 15-20 are based on the following passage:

(1) While all dogs descend through gray wolves, it's easy to notice that dog breeds come in a variety of shapes and sizes. (2) With such a drastic range of traits, appearances and body types, dogs are one of the most variable and adaptable species on the planet. (3) But why so many differences. (4) The answer is that humans have actually played a major role in altering the biology of dogs. (5) This was done through a process called selective breeding.

(6) Selective breeding which is also called artificial selection is the process in which animals with desired traits are bred in order to produce offspring that share the same traits. (7) In natural evolution, animals must adopt to their environments to increase their chance of survival. (8) Over time, certain traits develop in animals that enable them to thrive in these environments. (9) Those animals with more of these traits, or better versions of these traits, gain an advantage over others of their species. (10) Therefore, the animal's chances to mate are increased and these useful genes are passed into their offspring. (11) With dog breeding, humans select traits that are desired and encourage more of these desired traits in other dogs by breeding dogs that already have them.

15. Which sentence in the first paragraph uses an incorrect preposition?
    a. Sentence 1
    b. Sentence 2
    c. Sentence 4
    d. Sentence 5

16. Which sentence in the passage has a comma error?
    a. Sentence 2
    b. Sentence 6
    c. Sentence 9
    d. Sentence 11

17. Which sentence is missing an Oxford comma?
    a. Sentence 2
    b. Sentence 9
    c. Sentence 10
    d. Sentence 11

18. Which word in the second paragraph is incorrect?
    a. Offspring
    b. Adopt
    c. Traits
    d. Species

19. Which sentence has an end punctuation error?
    a. Sentence 2
    b. Sentence 3
    c. Sentence 6
    d. Sentence 10

20. Which sentence in the second paragraph uses the incorrect preposition?
    a. Sentence 6
    b. Sentence 9
    c. Sentence 10
    d. Sentence 11

Questions 21–25 are based on the following passage:

(1) We live in a savage world; that's just a simple fact. (2) It is a time of violence, when the need for self-defense is imperative. (3) Martial arts, like Ju-jitsu, still play a vital role in ones survival. (4) Ju-jitsu, however doesn't justify kicking people around, even when being harassed or attacked. (5) Today, laws prohibit the use of unnecessary force in self-defense; these serve to eliminate beating someone to a pulp once they have been neutralized. (6) Such laws are needed. (7) Apart from being unnecessary to continually strike a person when their down, its immoral. (8) Such over-aggressive retaliation turns the innocent into the aggressor. (9) Ju-jitsu provides a way for defending oneself while maintaining the philosophy of restraint and self-discipline. (10) Ingratiated into its core philosophy, Ju-jitsu tempers the potential to do great physical harm with respect for that power and for life.

21. Which of the following would be the best choice for Sentence3 (reproduced below)?

Martial arts, like Ju-jitsu, still play a vital role in ones survival.

    a. Leave it as is
    b. Martial arts, like Ju-jitsu, still play a vital role in one's survival.
    c. Martial arts, like Ju-jitsu still play a vital role in ones survival.
    d. Martial arts, like Ju-jitsu, still plays a vital role in one's survival.

22. Which of the following would be the best choice for the underlined portion of Sentence 4 (reproduced below)?

Ju-jitsu, however doesn't justify kicking people around, even when being harassed or attacked.

    a. Leave it as is
    b. Ju-jitsu, however, isn't justified by kicking people around,
    c. However, Ju-jitsu doesn't justify kicking people around,
    d. Ju-jitsu however doesn't justify kicking people around,

268

23. Which of the following would be the best choice for the underlined portion of Sentence 5 (reproduced below)?

Today, laws prohibit the <u>use of unnecessary force in self-defense; these serve to eliminate</u> beating someone to a pulp once they have been neutralized.

a. Leave it as is
b. use of unnecessary force in self-defense serving to eliminate
c. use of unnecessary force, in self-defense, these serve to eliminate
d. use of unnecessary force. In self-defense, these serve to eliminate

24. Which of the following would be the best choice for the underlined portion of Sentence 7 (reproduced below)?

Apart from being unnecessary to continually strike a person when <u>their down, its immoral.</u>

a. Leave it as is
b. their down, it's immoral.
c. they're down, its immoral.
d. they're down, it's immoral.

25. Which of the following would be the best choice for the underline portion of Sentence 10 (reproduced below)?

<u>Ingratiated into its core philosophy,</u> Ju-jitsu tempers the potential to do great physical harm with respect for that power, and for life.

a. Leave it as is
b. Ingratiated into its core philosophy,
c. Ingratiated into it's core philosophy,
d. Ingratiated into its' core philosophy,

Written Essay

For-profit institutions of learning should be illegal because they suffer from a conflict of interest between the student receiving the best education and the institution minimizing costs in providing instruction and other services to remain profitable.

Write a response in support or nonsupport of the statement in regard to the position that is taken. To support your stance, evaluate how your position may or may not be true to the argument. Explain how you've come to your position and provide examples in support

Answer Explanations #1

Arithmetic

1. C: The decimal points are lined up, with zeroes put in as needed. Then, the numbers are added just like integers:

$$
\begin{array}{r}
3.40 \\
2.35 \\
+4.00 \\
\hline
9.75
\end{array}
$$

2. A: This problem can be multiplied as 588×32, except at the end, the decimal point needs to be moved three places to the left. Performing the multiplication will give 18,816, and moving the decimal place over three places results in 18.816.

3. D: The fraction is converted so that the denominator is 100 by multiplying the numerator and denominator by 4, to get $\frac{3}{25} = \frac{12}{100}$. Dividing a number by 100 just moves the decimal point two places to the left, with a result of 0.12.

4. A: Figure out which is largest by looking at the first non-zero digits. Choice *B*'s first non-zero digit is in the hundredths place. The other three all have non-zero digits in the tenths place, so it must be *A*, *C*, or *D*. Of these, *A* has the largest first non-zero digit.

5. C: 40*N* would be 4000% of *N*. It's possible to check that each of the others is actually 40% of *N*.

6. B: Instead of multiplying these out, the product can be estimated by using $18 \times 10 = 180$. The error here should be lower than 15, since it is rounded to the nearest integer, and the numbers add to something less than 30.

7. C: 85% of a number means multiplying that number by 0.85. So, $0.85 \times 20 = \frac{85}{100} \times \frac{20}{1}$, which can be simplified to:

$$
\frac{17}{20} \times \frac{20}{1} = 17
$$

8. A: To find the fraction of the bill that the first three people pay, the fractions need to be added, which means finding the common denominator. The common denominator will be 60.

$$
\frac{1}{5} + \frac{1}{4} + \frac{1}{3}
$$

$$
\frac{12}{60} + \frac{15}{60} + \frac{20}{60} = \frac{47}{60}
$$

The remainder of the bill is:

$$
1 - \frac{47}{60} = \frac{60}{60} - \frac{47}{60} = \frac{13}{60}
$$

9. B: 30% is 3/10. The number itself must be 10/3 of 6, or:

$$\frac{10}{3} \times 6 = 10 \times 2 = 20$$

10. A: First, convert the mixed numbers to improper fractions: $\frac{11}{3} - \frac{9}{5}$. Then, use 15 as a common denominator:

$$\frac{11}{3} - \frac{9}{5}$$

$$\frac{55}{15} - \frac{27}{15} = \frac{28}{15} = 1\frac{13}{15}$$

11. A: Dividing by 98 can be approximated by dividing by 100, which would mean shifting the decimal point of the numerator to the left by 2. The result is 4.2 and rounds to 4.

12. B:

$$4\frac{1}{3} + 3\frac{3}{4} = 4 + 3 + \frac{1}{3} + \frac{3}{4}$$

$$7 + \frac{1}{3} + \frac{3}{4}$$

Adding the fractions gives:

$$\frac{1}{3} + \frac{3}{4} = \frac{4}{12} + \frac{9}{12}$$

$$\frac{13}{12} = 1 + \frac{1}{12}$$

Thus:

$$7 + \frac{1}{3} + \frac{3}{4}$$

$$7 + 1 + \frac{1}{12} = 8\frac{1}{12}$$

13. C: The average is calculated by adding all six numbers, then dividing by 6. The first five numbers have a sum of 25. If the total divided by 6 is equal to 6, then the total itself must be 36. The sixth number must be 36 − 25 = 11.

14. B: Multiplying by 10^{-3} means moving the decimal point three places to the left, putting in zeroes as necessary.

15. B: $\frac{5}{2} \div \frac{1}{3} = \frac{5}{2} \times \frac{3}{1} = \frac{15}{2} = 7.5$

16. A: The total fraction taken up by green and red shirts will be:

$$\frac{1}{3} + \frac{2}{5} = \frac{5}{15} + \frac{6}{15} = \frac{11}{15}$$

The remaining fraction is:

$$1 - \frac{11}{15} = \frac{15}{15} - \frac{11}{15} = \frac{4}{15}$$

17. C: If she has used 1/3 of the paint, she has 2/3 remaining. $2\frac{1}{2}$ gallons are the same as $\frac{5}{2}$ gallons. The calculation is:

$$\frac{2}{3} \times \frac{5}{2} = \frac{5}{3} = 1\frac{2}{3} \text{ gallons}$$

18. D: This problem can be solved by using unit conversions. The initial units are miles per minute. The final units need to be feet per second. Converting miles to feet uses the equivalence statement $1\ mile = 5{,}280\ feet$. Converting minutes to seconds uses the equivalence statement $1\ minute = 60\ seconds$. Setting up the ratios to convert the units is shown in the following equation: $\frac{72\ miles}{90\ minutes} \times \frac{1\ minute}{60\ seconds} \times \frac{5280\ feet}{1\ mile} = 70.4\ feet\ per\ second$. The initial units cancel out, and the new, desired units are left.

19. C: Scientific notation division can be solved by grouping the first terms together and grouping the tens together. The first terms can be divided, and the tens terms can be simplified using the rules for exponents. The initial expression becomes 0.4×10^4. This is not in scientific notation because the first number is not between 1 and 10. Shifting the decimal and subtracting one from the exponent, the answer becomes 4.0×10^3.

20. A: The relative error can be found by finding the absolute error and making it a percent of the true value. The absolute error is $36 - 35.75 = 0.25$. This error is then divided by 35.75—the true value—to find 0.7%.

Quantitative Reasoning, Algebra, and Statistics

1. B: To simplify the given equation, the first step is to make all exponents positive by moving them to the opposite place in the fraction. This expression becomes $\frac{4b^3b^2}{a^1a^4} \times \frac{3a}{b}$. Then the rules for exponents can be used to simplify. Multiplying the same bases means the exponents can be added. Dividing the same bases means the exponents are subtracted. Thus, after multiplying the exponents in the first fraction the equation becomes $\frac{4b^5}{a^5} \times \frac{3a}{b}$. Then, the two fractions can be multiplied, and the exponents must be divided to find $\frac{12b^4}{a^4}$.

2. D: The expression is three times the sum of twice a number and 1, which is $3(2x + 1)$. Then, 6 is subtracted from this expression.

3. A: Robert accomplished his task on Tuesday in $\frac{3}{4}$ the time compared to Monday. He must have worked $\frac{4}{3}$ as fast.

4. D: The slope from this equation is 50, and it is interpreted as the cost per gigabyte used. Since the g-value represents number of gigabytes and the equation is set equal to the cost in dollars, the slope relates these two values. For every gigabyte used on the phone, the bill goes up 50 dollars.

5. C: Because the triangles are similar, the lengths of the corresponding sides are proportional. Therefore:

$$\frac{30 + x}{30} = \frac{22}{14} = \frac{y + 15}{y}$$

This results in the equation $14(30 + x) = 22 \times 30$ which, when solved, gives $x = 17.1$. The proportion also results in the equation $14(y + 15) = 22y$ which, when solved, gives $y = 26.3$.

6. C: The numbers needed are those that add to -2 and multiply to -8. The difference between 2 and 4 is 2. Their product is 8, and -4 and 2 will work. Therefore:

$$x^2 - 2x - 8 = (x - 4)(x + 2)$$

The latter has roots 4 and -2 or 1 ± 3.

7. B: To factor $x^2 + 4x + 4$, the numbers needed are those that add to 4 and multiply to 4. Therefore, both numbers must be 2, and the expression factors to:

$$x^2 + 4x + 4 = (x + 2)^2$$

Similarly, the second expression factors to $x^2 - x - 6 = (x - 3)(x + 2)$, so that they have $x + 2$ in common.

8. D: This system of equations involves one quadratic function and one linear function, as seen from the degree of each equation. One way to solve this is through substitution. Solving for y in the second equation yields $y = x + 2$. Plugging this equation in for the y of the quadratic equation yields $x^2 - 2x + x + 2 = 8$. Simplifying the equation, it becomes $x^2 - x + 2 = 8$. Setting this equal to zero and factoring, it becomes $x^2 - x - 6 = 0 = (x - 3)(x + 2)$. Solving these two factors for x gives the zeros $x = 3, -2$. To find the y-value for the point, each number can be plugged in to either original equation. Solving each one for y yields the points $(3, 5)$ and $(-2, 0)$.

9. B: The technique of completing the square must be used to change $4x^2 + 4y^2 - 16x - 24y + 51 = 0$ into the standard equation of a circle. First, the constant must be moved to the right-hand side of the equals sign, and each term must be divided by the coefficient of the x^2 term (which is 4). The x and y terms must be grouped together to obtain $x^2 - 4x + y^2 - 6y = -\frac{51}{4}$. Then, the process of completing the square must be completed for each variable. This gives $(x^2 - 4x + 4) + (y^2 - 6y + 9) = -\frac{51}{4} + 4 + 9$. The equation can be written as $(x - 2)^2 + (y - 3)^2 = \frac{1}{4}$. Therefore, the center of the circle is $(2, 3)$ and the radius is $\sqrt{1/4} = 1/2$.

10. D: The midpoint formula should be used to get the average of both points.

$$M = \left(\frac{x_1 + x_2}{2}, \frac{y_1 + y_2}{2}\right) = \left(\frac{-1 + 3}{2}, \frac{2 + (-6)}{2}\right) = (1, -2)$$

11. A: First, the sample mean must be calculated. $\bar{x} = \frac{1}{4}(1 + 3 + 5 + 7) = 4$. The sample standard deviation of the data set is:

$$s = \sqrt{\frac{\sum(x - \bar{x})^2}{n - 1}}$$

and $n = 4$ represents the number of data points.

Therefore, the sample standard deviation is:

$$s = \sqrt{\frac{1}{3}[(1 - 4)^2 + (3 - 4)^2 + (5 - 4)^2 + (7 - 4)^2]}$$

$$s = \sqrt{\frac{1}{3}(9 + 1 + 1 + 9)} = 2.58$$

12. C: The sample space is made up of $8 + 7 + 6 + 5 = 26$ balls. The probability of pulling each individual ball is $1/26$. Since there are 7 yellow balls, the probability of pulling a yellow ball is $7/26$.

13. B: For the first card drawn, the probability of a King being pulled is $\frac{4}{52}$. Since this card isn't replaced, if a King is drawn first the probability of a King being drawn second is $\frac{3}{51}$. The probability of a King being drawn in both the first and second draw is the product of the two probabilities: $\frac{4}{52} \times \frac{3}{51} = \frac{12}{2652}$. This fraction, when divided by 12, equals $\frac{1}{221}$.

14. D: The addition rule is necessary to determine the probability because a 6 can be rolled on either roll of the die. The rule used is $P(A \text{ or } B) = P(A) + P(B) - P(A \text{ and } B)$. The probability of a 6 being individually rolled is $\frac{1}{6}$ and the probability of a 6 being rolled twice is $\frac{1}{6} \times \frac{1}{6} = \frac{1}{36}$. Therefore, the probability that a 6 is rolled at least once is $\frac{1}{6} + \frac{1}{6} - \frac{1}{36} = \frac{11}{36}$.

15. D: The slope is given by the change in y divided by the change in x. The change in y is 2-0 = 2, and the change in x is:

$$0 - (-4) = 4$$

The slope is $\frac{2}{4} = \frac{1}{2}$.

16. C: Janice will be choosing 4 employees out of a set of 6 applicants, so this will be given by the choice function. The following equation shows the choice function worked out:

$$\binom{6}{4} = \frac{6!}{4!\,(6 - 4)!} = \frac{6!}{4!\,(2)!}$$

$$\frac{6 \cdot 5 \cdot 4 \cdot 3 \cdot 2 \cdot 1}{4 \cdot 3 \cdot 2 \cdot 1 \cdot 2 \cdot 1} = \frac{6 \cdot 5}{2} = 15$$

17. C: The $(3x + 1)$ can be factored to get $(2x - 5)(3x + 1)$.

18. D: Because the coefficient of x^2 is negative, this function has a graph that is a parabola that opens downward. Therefore, it will be greater than 0 between its real roots, if it has any. Checking the discriminant, the result is:

$$1^2 - 4(-3)(-8) = 1 - 96 = -95$$

Since the discriminant is negative, this equation has no real solutions. Since this has no real roots, it must be always positive or always negative. Its graph opens downward, so it has at least some negative values. That means it is always negative. Thus, it is greater than zero for no real numbers.

19. A: Check each value, but it is easiest to use the quadratic formula, which gives

$$x = \frac{2 \pm \sqrt{(-2)^2 - 4(1)(-2)}}{2} = 1 \pm \frac{\sqrt{12}}{2}$$

$$1 \pm \frac{2\sqrt{3}}{2} = 1 \pm \sqrt{3}$$

The only one of these which appears as an answer choice is $1 + \sqrt{3}$.

20. B: The y coordinate of every point on the graph of $y = x^2 + 2$ has a vertex at (0,2) on the y-axis. The circle with a center at (0,1) also lies on the y-axis. With a radius of 1, the circle touches the parabola at one point. The vertex of the parabola (0,2).

Advanced Algebra and Functions

1. B: Start with the original equation: x- 2xy + 2y, then replace each instance of x with a 2, and each instance of y with a 3 to get:

$$2^2 - 2 \cdot 2 \cdot 3 + 2 \cdot 3^2 = 4 - 12 + 18 = 10$$

2. A: To expand a squared binomial, it's necessary to use the *First, Inner, Outer, Last Method.*

$$(2x - 4y)^2$$

$$(2x)(2x) + (2x)(-4y) + (-4y)(2x) + (-4y)(-4y)$$

$$4x^2 - 8xy - 8xy + 16y^2$$

$$4x^2 - 16xy + 16y^2$$

3. D: Factor the numerator into $x^2 - 6x + 9 = (x - 3)^2$, since:

$$-3 - 3 = -6$$

$$(-3)(-3) = 9$$

Factor the denominator into $x^2 - x - 6 = (x - 3)(x + 2)$, since:

$$-3 + 2 = -1,$$

$$(-3)(2) = -6$$

This means the rational function can be rewritten as:

$$\frac{x^2 - 6x + 9}{x^2 - x - 6} = \frac{(x-3)^2}{(x-3)(x+2)}$$

Using the restriction of x > 3, do not worry about any of these terms being 0, and cancel an $x - 3$ from the numerator and the denominator, leaving $\frac{x-3}{x+2}$.

4. C: The area of the shaded region is the area of the square, minus the area of the circle. The area of the circle will be πr^2. The side of the square will be $2r$, so the area of the square will be $4r^2$. Therefore, the difference is:

$$4r^2 - \pi r^2 = (4 - \pi)r^2$$

5. B: Add 3 to both sides to get $4x = 8$. Then divide both sides by 4 to get $x = 2$.

6. B: There are two zeros for the given function. They are $x = 0, -2$. The zeros can be found several ways, but this particular equation can be factored into $f(x) = x(x^2 + 4x + 4) = x(x + 2)(x + 2)$. By setting each factor equal to zero and solving for x, there are two solutions. On a graph, these zeros can be seen where the line crosses the x-axis.

7. D: This problem involves a composition function, where one function is plugged into the other function. In this case, the $f(x)$ function is plugged into the $g(x)$ function for each x-value. The composition equation becomes $g(f(x)) = 2^3 - 3(2^2) - 2(2) + 6$. Simplifying the equation gives the answer $g(f(x)) = 8 - 3(4) - 2(2) + 6 = 8 - 12 - 4 + 6 = -2$.

8. C: Graphing the function $y = \cos(x)$ shows that the curve starts at $(0, 1)$, has an amplitude of 2, and a period of 2π. This same curve can be constructed using the sine graph, by shifting the graph to the left $\frac{\pi}{2}$ units. This equation is in the form $y = \sin(x + \frac{\pi}{2})$.

9. D: Using SOHCAHTOA, tangent is $\frac{y}{x}$ for the special triangles. Since the value of $\tan(x)$ needs to be negative one, the angle for the tangent must be some form of 45 degrees or $\frac{\pi}{4}$. The value is negative in the second and fourth quadrant, so the answer is $\frac{3\pi}{4}$ and $\frac{7\pi}{4}$.

10. B: To simplify this inequality, subtract 3 from both sides to get:

$$-\frac{1}{2}x \geq -1$$

Then, multiply both sides by -2 (remembering this flips the direction of the inequality) to get $x \leq 2$.

11. D: There are two ways to approach this problem. Each value can be substituted into each equation. A can be eliminated, since:

$$4^2 + 16 = 32$$

Choice B can be eliminated, since:

$$4^2 + 4 \cdot 4 - 4 = 28$$

C can be eliminated, since

$$4^2 - 2 \cdot 4 - 2 = 6$$

But, plugging in either value into $x^2 - 16$, which gives:

$$(\pm 4)^2 - 16 = 16 - 16 = 0$$

12. A: The common denominator here will be $4x$. Rewrite these fractions as:

$$\frac{3}{x} + \frac{5u}{2x} - \frac{u}{4}$$

$$\frac{12}{4x} + \frac{10u}{4x} - \frac{ux}{4x}$$

$$\frac{12x + 10u - ux}{4x}$$

13. D: The expression is simplified by collecting like terms. Terms with the same variable and exponent are like terms, and their coefficients can be added.

14. D: Finding the zeros for a function by factoring is done by setting the equation equal to zero, then completely factoring. Since there was a common x for each term in the provided equation, that would be factored out first. Then the quadratic that was left could be factored into two binomials, which are $(x + 1)(x - 4)$. Setting each factor equal to zero and solving for x yields three zeros.

15. D: Dividing rational expressions follows the same rule as dividing fractions. The division is changed to multiplication by the reciprocal of the second fraction. This turns the expression into:

$$\frac{5x^3}{3x^2y} \times \frac{3y^9}{25}$$

Multiplying across and simplifying, the final expression is $\frac{xy^8}{5}$.

16. B: The table shows values that are increasing exponentially. The differences between the inputs are the same, while the differences in the outputs are changing by a factor of 2. The values in the table can be modeled by the equation $f(x) = 2^x$.

17. A: Using the trigonometric identity $\tan(\theta) = \frac{\sin(\theta)}{\cos(\theta)}$, the expression becomes $\frac{\sin \theta}{\cos \theta} \cos \theta$. The factors that are the same on the top and bottom cancel out, leaving the simplified expression $\sin \theta$.

18. C: The equation $x = 3$ is not a function because it does not pass the vertical line test. This test is made from the definition of a function, where each x-value must be mapped to one and only one y-value. This equation is a vertical line, so the x-value of 3 is mapped with an infinite number of y-values.

19. A: The function can be factored to identify the zeros. First, the term $3x$ is factored out to the front because each term contains $3x$. Then, the quadratic is factored into $(x + 3)(x - 2)$.

20. B: Plugging into the quadratic formula yields, for solutions:

$$x = \frac{1 \pm \sqrt{-1 + 4 \cdot 1(-3)}}{2} = \frac{1}{2} \pm \frac{\sqrt{11}}{2}$$

Therefore:

$$x - \frac{1}{2} = \pm \frac{\sqrt{11}}{2}$$

Now, if this is squared, then the $\pm$ cancels and left with:

$$\left(\frac{\sqrt{11}}{2}\right)^2 = \frac{11}{4}$$

Reading Comprehension

1. A: The tone is exasperated. While contemplative is an option because of the inquisitive nature of the text, Choice *A* is correct because the speaker is frustrated by the thought of being included when he felt that the fellow members of his race were being excluded. The speaker is not nonchalant, nor accepting of the circumstances which he describes.

2. C: Choice *C*, *contented*, is the only word that has different meaning. Furthermore, the speaker expresses objection and disdain throughout the entire text.

3. B: To address the hypocrisy of the Fourth of July holiday. While the speaker makes biblical references, it is not the main focus of the passage, thus eliminating Choice *A* as an answer. The passage also makes no mention of wealthy landowners and doesn't speak of any positive response to the historical events, so Choices *C* and *D* are not correct.

4. D: Choice *D* is the correct answer because it clearly makes reference to justice being denied.

5. D: Hyperbole. Choices *A* and *B* are unrelated. Assonance is the repetition of sounds and commonly occurs in poetry. Parallelism refers to two statements that correlate in some manner. Choice *C* is incorrect because amplification normally refers to clarification of meaning by broadening the sentence structure, while hyperbole refers to a phrase or statement that is being exaggerated.

6. C: Display the equivocation of the speaker and those that he represents. Choice *C* is correct because the speaker is clear about his intention and stance throughout the text. Choice *A* could be true, but the words "common text" is arguable. Choice *B* is also partially true, as another group of people affected by slavery are being referenced. However, the speaker is not trying to convince the audience that injustices have been committed, as it is already understood there have been injustices committed. Choice *D* is also close to the correct answer, but it is not the *best* answer choice possible.

7. A: The passage describes a situation and then explains the causes that led to it. Also, it utilizes cause and effect signal words, such as *causes, factors, so,* and *as a result. B* is incorrect because a compare and contrast order considers the similarities and differences of two or more things. *C* is incorrect because spatial order describes where things are located in relation to each other. Finally, *D* is incorrect because time order describes when things occurred chronologically.

8. C: One can reasonably infer that Gulliver is considerably larger than the children who were playing around him because multiple children could fit into his hand. Choice *B* is incorrect because there is no indication of stress in Gulliver's tone. Choices *A* and *D* aren't the best answer because though Gulliver seems fond of his new acquaintances, he didn't travel there with the intentions of meeting new people or to express a definite love for them in this particular portion of the text.

9. D: Choice *D* looks like a strong answer. This answer choice references the argument's main points—*Infinite Jest* is a modern classic, the book deserves its praise, and everybody should read it. In contrast, Choice *A* restates the author's conclusion. The correct answer to main point questions will often be closely related to the conclusion. Choice *B* restates a premise. Is the author's main point that *Infinite Jest* is a page-turner? No, he uses readers' obsession with the book as a premise. Eliminate this choice. Choice *C* is definitely not the main point of the passage. It's a simple fact underlying the argument. It certainly cannot be considered the main point. Eliminate this choice.

10. A: Choice *A* is consistent with the argument's logic. The argument asserts that the world powers' military alliances amounted to a lit fuse, and the assassination merely lit it. The main point of the argument is that any event involving the military alliances would have led to a world war. Choice *B* runs counter to the argument's tone and reasoning. It can immediately be eliminated. Choice *C* is also clearly incorrect. At no point does the argument blame any single or group of countries for starting World War I. Choice *D* is incorrect for the same reason as Choice *C*. Eliminate this choice.

11. A: The purpose is to inform the reader about what assault is and how it is committed. Choice *B* is incorrect because the passage does not state that assault is a lesser form of lethal force, only that an assault can use lethal force, or alternatively, lethal force can be utilized to counter a dangerous assault. Choice *C* is incorrect because the passage is informative and does not have a set agenda. Finally, Choice *D* is incorrect because although the author uses an example in order to explain assault, it is not indicated that this is the author's personal account.

12. C: The situation of the man who is attacked in an alley by another man with a knife would most merit the use of lethal force. If the man being attacked used self-defense by lethal force, it would not be considered illegal. The presence of a deadly weapon indicates mal-intent and because the individual is isolated in an alley, lethal force in self-defense may be the only way to preserve his life. Choices *A* and *B* can be ruled out because in these situations, no one is in danger of immediate death or bodily harm by someone else. Choice *D* is an assault that does exhibit intent to harm, but this situation isn't severe enough to merit lethal force; there is no intent to kill.

13. B: As discussed in the second passage, there are several forms of assault, like assault with a deadly weapon, verbal assault, or threatening posture or language. Choice *A* is incorrect because lethal force and assault are separate as indicated by the passages. Choice *C* is incorrect because assault is never justified. Self-defense resulting in lethal force can be justified. Choice *D* is incorrect because the author does mention what the charges are on assaults; therefore, we cannot assume that they are more or less than unnecessary use of force charges.

14. D: The use of lethal force is not evaluated on the intent of the user but rather the severity of the primary attack that warranted self-defense. This statement most undermines the last part of the passage because it directly contradicts how the law evaluates the use of lethal force. Choices *A* and *B* are stated in the paragraph, and therefore do not undermine the explanation from the author. Choice *C* does not necessarily undermine the passage, but it does not support the passage either. It is more of an opinion that does not strengthen or weaken the explanation.

15. C: An assault with deadly intent can lead to an individual using lethal force to preserve their well-being. Choice *C* is correct because it clearly establishes what both assault and lethal force are and gives the specific way in which the two concepts meet. Choice *A* is incorrect because lethal force doesn't necessarily result in assault. Choice *B* is incorrect because it contradicts the information in the passage (that assault with deadly intent can lead to an individual using lethal force). Choice *D* is compelling but ultimately too vague; the statement touches on aspects of the two ideas but fails to present the concrete way in which the two are connected to each other.

16. D: A *rookery* is a colony of breeding birds. Although *rookery* could mean Choice *A*, houses in a slum area, it does not make sense in this context. Choices *B* and *C* are both incorrect, as this is not a place for hunters to trade tools or for wardens to trade stories.

17. B: An important bird colony. The previous sentence is describing "twenty colonies" of birds, so what follows should be a bird colony. Choice *A* may be true, but we have no evidence of this in the text. Choice *C* does touch on the tension between the hunters and wardens, but there is no official "Bird Island Battle" mentioned in the text. Choice *D* does not exist in the text.

18. D: To demonstrate the success of the protective work of the Audubon Association. The text mentions several different times how and why the association has been successful and gives examples to back this fact. Choice *A* is incorrect because although the article, in some instances, calls certain people to act, it is not the purpose of the entire passage. There is no way to tell if Choices *B* and *C* are correct, as they are not mentioned in the text.

19 C: To have a better opportunity to hunt the birds. Choice *A* might be true in a general sense, but it is not relevant to the context of the text. Choice *B* is incorrect because the hunters are not studying lines of flight to help wardens, but to hunt birds. Choice *D* is incorrect because nothing in the text mentions that hunters are trying to build homes underneath lines of flight of birds for good luck.

20. A: It introduces certain insects that transition from water to air. Choice *B* is incorrect because although the passage talks about gills, it is not the central idea of the passage. Choices *C* and *D* are incorrect because the passage does not "define" or "invite," but only serves as an introduction to stoneflies, dragonflies, and mayflies and their transition from water to air.

Writing

1. B: Choice *B* is correct because it removes the apostrophe from *icon's*, since the noun *icon* is not possessing anything. This conveys the author's intent of setting *Frankenstein* apart from other icons of the romantic and science fiction genres. Choices *A* and *C* are therefore incorrect. Choice *D* is a good revision but alters the meaning of the sentence—*Frankenstein* is one of the icons, not the sole icon.

2 C: Choice *C* correctly adds a comma after *style*, successfully joining the dependent and the independent clauses as a single sentence. Choice *A* is incorrect because the dependent and independent clauses remain unsuccessfully combined without the comma. Choices *B* and *D* do nothing to fix this.

3. A: Choice *A* is correct, as the sentence doesn't require changes. Choice *B* incorrectly changes the noun *enlightenment* into the verb *enlighten*. Choices *C* and *D* alter the original meaning of the sentence.

4. B: Choice *B* is correct, fixing the incorrect split of *highlight*. This is a polyseme, a word combined from two unrelated words to make a new word. On their own, *high* and *light* make no sense for the sentence, making Choice *A* incorrect. Choice *C* incorrectly decapitalizes *Ancient*—since it modifies *Greece* and

works with the noun to describe a civilization, *Ancient Greece* functions as a proper noun, which should be capitalized. Choice *D* uses *highlighting*, a gerund, but the present tense of *highlight* is what works with the rest of the sentence; to make this change, a comma would be needed after *Rome*.

5. A: Choice *A* is correct, as *not only* and *but also* are correlative pairs. In this sentence, *but* successfully transitions the first part into the second half, making punctuation unnecessary. Additionally, the use of *to* indicates that an idea or challenge is being presented to the reader. Choices *B*, *C*, and *D* are not as active, meaning these revisions weaken the sentence.

6. D: Choice *D* is correct, adding finer details to help the reader understand exactly what Prometheus did and his impact: fire came with knowledge and power. Choice *A* lacks a comma after *fire*. Choice *B* inserts unnecessary commas since *people* is not part of the list *knowledge and power*. Choice *C* is a strong revision but could be confusing, hinting that the fire was knowledge and power itself, as opposed to being symbolized by the fire.

7. C: Choice *C* reverses the order of the section, making the sentence more direct. Choice *A* lacks a comma after *gods*, and although Choice *B* adds this, the structure is too different from the first half of the sentence to flow correctly. Choice *D* is overly complicated and repetitious in its structure even though it doesn't need any punctuation.

8. B: Choice *B* is the correct answer because it states a general fact about Japanese food and serves as a main idea for the paragraph. Although Choices *A, C,* and *D* are true facts about Japanese cuisine, they are details that would further support a main idea about Japanese food.

9. C: Choice *C* is the correct answer. Sentence 10 should be the next sentence after sentence 8. It introduces the town of Hakone and provides the framework for the specific information about the town presented in sentences 9 through 12.

10. D: Choice *D* is the correct answer. This sentence is more concise and still incorporates all of the information trying to be conveyed in the original sentence. Choices *A, B,* and *C* either mix up the information in the original sentence or give incorrect facts.

11. B: Choice *B* is the correct answer. In this sentence, the words *or bullet train* serve as an appositive, or extra information provided in the sentence. These words are added to the sentence to define *Shinkansen*. However, the sentence would be correct even if these words hadn't been added, so this phrase should be set apart with commas. A correct version of the sentence would look like this: [8]*People from all over Japan ride the Shinkansen, or bullet train, to Osaka to watch the events.*

12. A: Choice *A* is the correct answer. The transition phrase, *On this day*, helps to clarify that things are closed for a specific reason versus being closed in general.

13. A: Choice *A* is the correct answer. In sentence 1, the writer mentions Memorial Day and then Veterans Day in the statement of the main idea. Thus, the supporting sentences in the paragraph should also be in this order.

14. D: Choice *D* is the correct answer. This sentence rewords the writer's original sentence while still including all of the correct information. Choices *A, B,* and *C* either remain too wordy, provide inaccurate information, or fail to clarify the writer's idea.

15. A: Sentence 1 has the incorrect preposition because the word *through* is incorrectly used here. The correct preposition is *from* to describe the fact that dogs are related to wolves, so the sentence should

281

be: While all dogs descend *from* gray wolves, it's easy to notice that dog breeds come in a variety of shapes and sizes.

16. B: Sentence 6 is missing necessary commas. The sentence should be: Selective breeding, which is also called artificial selection, is the process in which animals with desired traits are bred in order to produce offspring that share the same traits. The added commas serve to distinguish that *artificial selection* is just another term for *selective breeding* before the sentence continues. The structure is preserved, and the sentence can flow with more clarity.

17. A: Sentence 2 lacks the Oxford Comma, which helps clearly separate specific terms. The sentence should be: With such a drastic range of traits, appearances, and body types, dogs are one of the most variable and adaptable species on the planet.

18: B: *Adopt* is not the correct term. The author should have written *adapt,* because he or she is talking about how animals must modify themselves in some way to accommodate the changes in their environment. *Adopt* is to take in something as your own.

19. B: Sentence 3 has an end punctuation error because questions do not end with periods. A question mark is necessary.

20. C: Sentence 10 contains the error because the use of *into* is inappropriate for this context. The sentence should have used *on to*, describing the way genes are passed generationally.

21. B: Choice *B* is correct because it adds an apostrophe to *ones*, which indicates *one's* possession of *survival*. Choice *A* doesn't do this, so it is incorrect. This is the same for Choice *C*, but that option also takes out the crucial comma after *Ju-jitsu*. Choice *D* is incorrect because it changes *play* to *plays*. This disagrees with the plural *Martial arts*, exemplified by having an example of its many forms, *Ju-jitsu*. Therefore, *play* is required.

22. C: Choice *C* is the best answer because it most clearly defines the point that the author is trying to make. The original sentence would need a comma after *however* in order to continue the sentence fluidly—but this option isn't available. Choice *B* is close, but this option changes the meaning of the sentence. Therefore, the best alternative is to begin the sentence with *However* and have a comma follow right after it in order to introduce a new idea. The original context is still maintained, but the flow of the language is more streamlined. Thus, Choice *A* is incorrect. Choice *D* would need a comma before and after *however*, so it is also incorrect.

23. A: Choice *A* is the best answer for several reasons. To begin, the section is grammatically correct in using a semicolon to connect the two independent clauses. This allows the two ideas to be connected without separating them. In this context, the semicolon makes more sense for the overall sentence structure and passage as a whole. Choice *B* is incorrect because it forms a run-on. Choice *C* applies a comma in incorrect positions. Choice *D* separates the sentence in a place that does not make sense for the context.

24. D: Choice *D* is the correct answer because it fixes two key issues. First, *their* is incorrectly used. *Their* is a possessive indefinite pronoun and also an antecedent—neither of these fit the context of the sentence, so Choices *A* and *B* are incorrect. What should be used instead is *they're*, which is the contraction of *they are*, emphasizing action or the result of action in this case. Choice *D* also corrects another contraction-related issue with *its*. Again, *its* indicates possession, while *it's* is the contraction of

it is. The latter is what's needed for the sentence to make sense and be grammatically correct. Thus, Choice *C* is also incorrect.

25. A: Choice *A* is correct because the section contains no errors and clearly communicates the writer's point. Choice *B* is incorrect because it lacks a comma after *philosophy*, needed to link the first clause with the second. Choice *C* also has this issue but additionally alters *its* to *it's*; since *it is* does not make sense in this sentence, this is incorrect. Choice *D* is incorrect because *its* is already plural possessive and does not need an apostrophe on the end.

Practice Test #2

Arithmetic

1. What is $\frac{660}{100}$ rounded to the nearest integer?
   a. 67
   b. 66
   c. 7
   d. 6

2. Which of the following is largest?
   a. -0.45
   b. -0.096
   c. -0.3
   d. -0.313

3. What is the value of b in this equation?

$$5b - 4 = 2b + 17$$

   a. 13
   b. 24
   c. 7
   d. 21

4. Katie works at a clothing company and sold 192 shirts over the weekend. $^1/_3$ of the shirts that were sold were patterned, and the rest were solid. Which mathematical expression would calculate the number of solid shirts Katie sold over the weekend?
   a. $192 \times \frac{1}{3}$
   b. $192 \div \frac{1}{3}$
   c. $192 \times (1 - \frac{1}{3})$
   d. $192 \div 3$

5. Arrange the following numbers from least to greatest value:
   $0.85, \frac{4}{5}, \frac{2}{3}, \frac{91}{100}$
   a. $0.85, \frac{4}{5}, \frac{2}{3}, \frac{91}{100}$
   b. $\frac{4}{5}, 0.85, \frac{91}{100}, \frac{2}{3}$
   c. $\frac{2}{3}, \frac{4}{5}, 0.85, \frac{91}{100}$
   d. $0.85, \frac{91}{100}, \frac{4}{5}, \frac{2}{3}$

6. Divide and reduce 5/13 ÷ 25/169.
    a. 13/5
    b. 65/25
    c. 25/65
    d. 5/13

7. $\frac{14}{15} + \frac{3}{5} - \frac{1}{30} =$

    a. $\frac{19}{15}$

    b. $\frac{43}{30}$

    c. $\frac{4}{3}$

    d. $\frac{3}{2}$

8. A piggy bank contains 12 dollars' worth of nickels. A nickel weighs 5 grams, and the empty piggy bank weighs 1050 grams. What is the total weight of the full piggy bank?
    a. 1,110 grams
    b. 1,200 grams
    c. 2,250 grams
    d. 2,200 grams

9. Last year, the New York City area received approximately $27\frac{3}{4}$ inches of snow. The Denver area received approximately 3 times as much snow as New York City. How much snow fell in Denver?
    a. 60 inches

    b. $27\frac{1}{4}$ inches

    c. $9\frac{1}{4}$ inches

    d. $83\frac{1}{4}$ inches

10. What is the solution to the following problem in decimal form?
$$\frac{3}{5} \times \frac{7}{10} \div \frac{1}{2}$$
    a. 0.042
    b. 84%
    c. 0.84
    d. 0.42

11. After a 20% discount, Frank purchased a new refrigerator for $850. How much did he save from the original price?
    a. $170
    b. $212.50
    c. $105.75
    d. $200

12. Jessica buys 10 cans of paint. Red paint costs $1 per can and blue paint costs $2 per can. In total, she spends $16. How many red cans did she buy?
    a. 2
    b. 3
    c. 4
    d. 5

13. If a car can travel 300 miles in 4 hours, how far can it go in an hour and a half?
    a. 100 miles
    b. 112.5 miles
    c. 135.5 miles
    d. 150 miles

14. At the store, Jan spends $90 on apples and oranges. Apples cost $1 each and oranges cost $2 each. If Jan buys the same number of apples as oranges, how many oranges did she buy?
    a. 20
    b. 25
    c. 30
    d. 35

15. A train traveling 50 miles per hour takes a trip lasting 3 hours. If a map has a scale of 1 inch per 10 miles, how many inches apart are the train's starting point and ending point on the map?
    a. 14
    b. 12
    c. 13
    d. 15

16. A traveler takes an hour to drive to a museum, spends 3 hours and 30 minutes there, and takes half an hour to drive home. What percentage of his or her time was spent driving?
    a. 15%
    b. 30%
    c. 40%
    d. 60%

17. Simplify the following fraction:

$$\frac{\frac{5}{7}}{\frac{9}{11}}$$

    a. $\frac{55}{63}$

    b. $\frac{7}{1000}$

    c. $\frac{13}{15}$

    d. $\frac{5}{11}$

18. Greg buys a $10 lunch with 5% sales tax. He leaves a $2 tip after his bill. How much money does he spend?

    a. $12.50

    b. $12

    c. $13

    d. $13.25

19. Marty wishes to save $150 over a 4-day period. How much must Marty save each day on average?

    a. $37.50

    b. $35

    c. $45.50

    d. $41

20. A couple buys a house for $150,000. They sell it for $165,000. By what percentage did the house's value increase?

    a. 10%

    b. 13%

    c. 15%

    d. 17%

Quantitative Reasoning, Algebra, and Statistics

1. What is the value of x in the following expression?

$$4x - 12 = -2x$$

    a. 2

    b. 3

    c. -6

    d. -2

2. What is the simplified quotient of the following expression?

$$\frac{5x^3}{3x^2y} \div \frac{25}{3y^9}$$

    a. $\frac{125x}{9y^{10}}$

    b. $\frac{x}{5y^8}$

    c. $\frac{5}{xy^8}$

    d. $\frac{xy^8}{5}$

3. The area of a given rectangle is 24 square centimeters. If the measure of each side is multiplied by 3, what is the area of the new figure?

    a. 48cm

    b. 72cm

    c. 216cm

    d. 13,824cm

7. The table below shows tickets purchased during the week for entry to the local zoo. What is the mean of adult tickets sold for the week?

Day of the Week	Age	Tickets Sold
Monday	Adult	22
Monday	Child	30
Tuesday	Adult	16
Tuesday	Child	15
Wednesday	Adult	24
Wednesday	Child	23
Thursday	Adult	19
Thursday	Child	26
Friday	Adult	29
Friday	Child	38

a. 24.2
b. 21
c. 22
d. 26.4

8. Courtney leaves home and drives her truck 1,236 yards towards her destination. Her destination is 6,292 feet away from her home. How many more feet does she need to travel before she arrives?
a. 2,284

b. 3,708

c. 5,056

d. 2,584

9. A local candy store reports that of the 100 customers that bought suckers, 35 of them bought cherry. What is the probability of selecting 2 customers simultaneously at random that both purchased a cherry sucker?

a. $\frac{119}{990}$

b. $\frac{35}{100}$

c. $\frac{49}{400}$

d. $\frac{69}{99}$

10. Which inequality represents the number line below?

a. $4x + 5 < 8$
b. $-4x + 5 < 8$
c. $-4x + 5 > 8$
d. $4x - 5 > 8$

11. Which of the following formulas would correctly calculate the perimeter of a legal-sized piece of paper that is 14 inches long and $8\frac{1}{2}$ inches wide?

    a. $P = 14 + 8\frac{1}{2}$

    b. $P = 14 + 8\frac{1}{2} + 14 + 8\frac{1}{2}$

    c. $P = 14 \times 8\frac{1}{2}$

    d. $P = 14 \times \frac{17}{2}$

12. What is the slope of the line that passes through the points $(10, -4)$ and $(-5, 8)$?

    a. $-\frac{5}{4}$

    b. $-\frac{4}{15}$

    c. $-\frac{4}{5}$

    d. $-\frac{12}{5}$

13. Points L and M lie on line KN. The length of line KN is 30 units long, LN is 20 units long, and KM is 16 units long. How many units long is LM?

    a. 16
    b. 14
    c. 10
    d. 6

14. Angle y measures 48°. Angle x measures twice the value of angle y. What is the value of angle z?

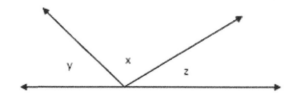

    a. 96°

    b. 36°

    c. 132°

    d. 42°

15. What is the measurement of angle f in the following picture? Assume the lines are parallel.

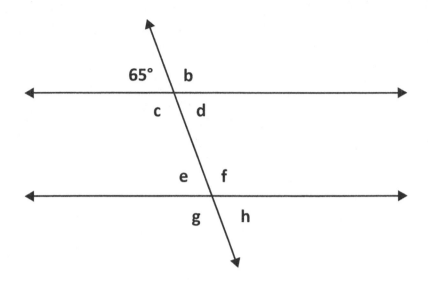

    a. 65 degrees
    b. 115 degrees
    c. 125 degrees
    d. 55 degrees

16. Given the triangle below, find the value of x if $y = 21$.

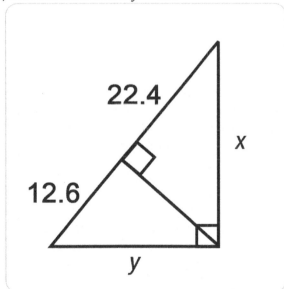

    a. 35
    b. 28
    c. 25
    d. 26

17. What is the value of x in the following triangle?

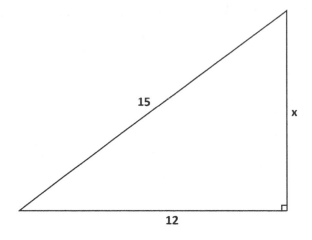

    a. 19.2
    b. 9
    c. 3
    d. 7.5

18. A data set is comprised of the following values: 30, 33, 33, 26, 27, 32, 33, 35, 29, 27. Which of the following has the greatest value?
    a. Mean
    b. Median
    c. Mode
    d. Range

19. A group of marathon runners were surveyed to see how many miles they ran in a week to prepare for the upcoming race. Six runners were surveyed, and the median of the responses was 14 miles. Which of the following is a possible list of responses?
    a. 11, 12, 15, 16, 11, 13
    b. 13, 16, 11, 12, 15, 16
    c. 16, 16, 14, 13, 11, 12
    d. 12, 15, 16, 13, 16, 15

20. 250 students were asked their favorite flavor of ice cream. The results are presented in the pie chart below. What is the probability of choosing a student who likes vanilla or strawberry ice cream to receive a free ice cream cone?

Favorite Ice Cream Flavor

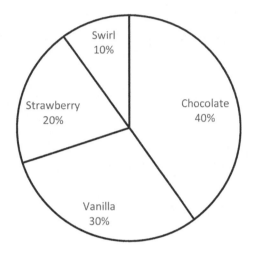

a. $\frac{3}{10}$

b. $\frac{2}{10}$

c. $\frac{1}{2}$

d. $\frac{1}{20}$

Advanced Algebra and Functions

1. If $x = 2.6$ and $y = 5.3$, what is the value of $3.2xy - 4.1y$?
   a. -7.95
   b. 33.436
   c. 44.096
   d. 22.366

2. What is the solution to the radical equation $\sqrt[3]{2x + 11} + 9 = 12$?
   a. -8
   b. 8
   c. 0
   d. 12

3. What is the solution to the following system of equations?
$$\begin{cases} x^2 + y = 4 \\ 2x + y = 1 \end{cases}$$

    a. (–1, 3)
    b. (–1, 3), (3, –5)
    c. (3, –5)
    d. –1, 3

4. If $f(x) = x^2 - 3x + 17$, then what is $f(x + 1)$?
    a. $x^2 - 3x + 19$
    b. $x^2 - x + 15$
    c. $x^2 + 2x + 18$
    d. $x^2 - 3x + 14$

5. What is the product of the following expression?
$$(x + 2)(x^2 + 5x - 6)$$

    a. $8x^2 + 4x - 12$
    b. $x^2 + 6x - 4$
    c. $x^3 + 7x^2 + 4x - 12$
    d. $x^3 + 5x^2 - 4x - 12$

6. What are the roots of $x^2 + x - 2$?
    a. 1 and -2
    b. -1 and 2
    c. 2 and -2
    d. 9 and 13

7. What is the y-intercept of $y = x^{5/3} + (x - 3)(x + 1)$?
    a. 3.5
    b. 7.6
    c. -3
    d. -15.1

8. $(4x^2 y^4)^{\frac{3}{2}}$ can be simplified to which of the following?
    a. $8x^3 y^6$

    b. $4x^{\frac{5}{2}} y$

    c. $4xy$

    d. $32x^{\frac{7}{2}} y^{\frac{11}{2}}$

9. Which graph will be a line parallel to the graph of $y = 3x - 2$?
    a. $2y - 6x = 2$
    b. $y - 4x = 4$
    c. $3y = x - 2$
    d. $2x - 2y = 2$

10. A company invests $50,000 in a building where they can produce saws. If the cost of producing one saw is $40, then which function expresses the amount of money the company pays? The variable y is the money paid and x is the number of saws produced.

    a. $y = 50{,}000x + 40$
    b. $y + 40 = x - 50{,}000$
    c. $y = 40x - 50{,}000$
    d. $y = 40x + 50{,}000$

11. If $3x = 6y = -2z = 24$, then what does $4xy + z$ equal?

    a. 116
    b. 130
    c. 84
    d. 108

12. Which inequality represents the following number line?

    a. $-\frac{5}{2} \leq x < \frac{3}{2}$

    b. $-\frac{7}{2} \leq x < \frac{5}{2}$

    c. $-\frac{5}{2} < x \leq \frac{3}{2}$

    d. $\frac{5}{2} < x \leq -\frac{3}{2}$

13. What is the equation of a circle whose center is (0, 0) and whole radius is 5?

    a. $(x - 5)^2 + (y - 5)^2 = 25$
    b. $(x)^2 + (y)^2 = 5$
    c. $(x)^2 + (y)^2 = 25$
    d. $(x + 5)^2 + (y + 5)^2 = 25$

14. Which of the ordered pairs below is a solution to the following system of inequalities?

$$y > 2x - 3$$
$$y < -4x + 8$$

    a. (4, 5)
    b. (-3, -2)
    c. (3, -1)
    d. (5, 2)

15. Which equation best represents the scatterplot below?

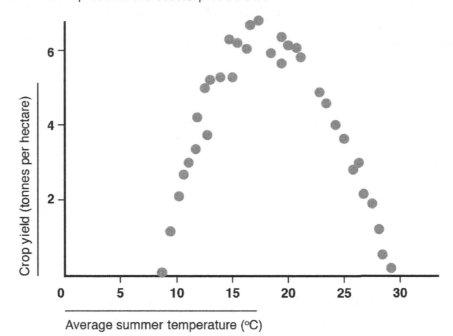

a. $y = 3x - 4$
b. $y = 2x^2 + 7x - 9$
c. $y = (3)(4^x)$
d. $y = -\frac{1}{14}x^2 + 2x - 8$

16. The triangle shown below is a right triangle. What's the value of x?

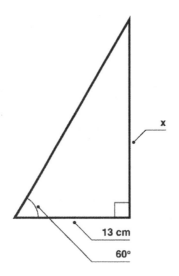

a. $x = 1.73$
b. $x = 0.57$
c. $x = 13$
d. $x = 22.49$

17. If $n = 2^2$, and $m = n^2$, then m^n equals?

    a. 2^{12}

    b. 2^{10}

    c. 2^{18}

    d. 2^{16}

18. The graph of the function $f(x) = |x + 3| - 4$ is graphed on the coordinate plane below.

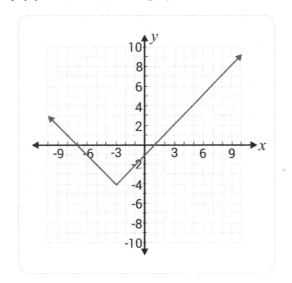

For which of the following functions does the graph of $g(x)$ intersect the function $f(x)$ exactly two times?

    a. $g(x) = x - 4$

    b. $g(x) = -x$

    c. $g(x) = x + 2$

    d. $g(x) = \frac{x}{2}$

19. If h is a multiple of 5, which of the following must also be multiples of 5?

    a. $5h - 3$

    b. $4h + 1$

    c. $10h + 20$

    d. $h^3 + 6$

20. Of the given sets of coordinates below, which ones lie on the line that is perpendicular to $y = 2x - 3$ and passes through the point $(0,5)$?

    a. $(2, 4)$

    b. $(-2, 7)$

    c. $(2, 6)$

    d. $(4, 3)$

    e. $(-6, 10)$

Reading Comprehension

Questions 1-6 are based on the following passage:

When researchers and engineers undertake a large-scale scientific project, they may end up making discoveries and developing technologies that have far wider uses than originally intended. This is especially true in NASA, one of the most influential and innovative scientific organizations in America. NASA spinoff technology refers to innovations originally developed for NASA space projects that are now used in a wide range of different commercial fields. Many consumers are unaware that products they are buying are based on NASA research! Spinoff technology proves that it is worthwhile to invest in science research because it could enrich people's lives in unexpected ways.

The first spinoff technology worth mentioning is baby food. In space, where astronauts have limited access to fresh food and fewer options about their daily meals, malnutrition is a serious concern. Consequently, NASA researchers were looking for ways to enhance the nutritional value of astronauts' food. Scientists found that a certain type of algae could be added to food, improving the food's neurological benefits. When experts in the commercial food industry learned of this algae's potential to boost brain health, they were quick to begin their own research. The nutritional substance from algae then developed into a product called life's DHA, which can be found in over 90% of infant food sold in America.

Another intriguing example of a spinoff technology can be found in fashion. People who are always dropping their sunglasses may have invested in a pair of sunglasses with scratch resistant lenses—that is, it's impossible to scratch the glass, even if the glasses are dropped on an abrasive surface. This innovation is incredibly advantageous for people who are clumsy, but most shoppers don't know that this technology was originally developed by NASA. Scientists first created scratch resistant glass to help protect costly and crucial equipment from getting scratched in space, especially the helmet visors in space suits. However, sunglasses companies later realized that this technology could be profitable for their products, and they licensed the technology from NASA.

1. What is the main purpose of this article?
   a. To advise consumers to do more research before making a purchase
   b. To persuade readers to support NASA research
   c. To tell a narrative about the history of space technology
   d. To define and describe instances of spinoff technology

2. What is the organizational structure of this article?
   a. A general definition followed by more specific examples
   b. A general opinion followed by supporting arguments
   c. An important moment in history followed by chronological details
   d. A popular misconception followed by counterevidence

3. Why did NASA scientists research algae?
   a. They already knew algae was healthy for babies.
   b. They were interested in how to grow food in space.
   c. They were looking for ways to add health benefits to food.
   d. They hoped to use it to protect expensive research equipment.

4. What does the word "neurological" mean in the second paragraph?
   a. Related to the body
   b. Related to the brain
   c. Related to vitamins
   d. Related to technology

5. Why does the author mention space suit helmets?
   a. To give an example of astronaut fashion
   b. To explain where sunglasses got their shape
   c. To explain how astronauts protect their eyes
   d. To give an example of valuable space equipment

6. Which statement would the author probably NOT agree with?
   a. Consumers don't always know the history of the products they are buying.
   b. Sometimes new innovations have unexpected applications.
   c. It is difficult to make money from scientific research.
   d. Space equipment is often very expensive.

Questions 7-13 are based on the following passage:

People who argue that William Shakespeare is not responsible for the plays attributed to his name are known as anti-Stratfordians (from the name of Shakespeare's birthplace, Stratford-upon-Avon). The most common anti-Stratfordian claim is that William Shakespeare simply was not educated enough or from a high enough social class to have written plays overflowing with references to such a wide range of subjects like history, the classics, religion, and international culture. William Shakespeare was the son of a glove-maker, he only had a basic grade school education, and he never set foot outside of England—so how could he have produced plays of such sophistication and imagination? How could he have written in such detail about historical figures and events, or about different cultures and locations around Europe? According to anti-Stratfordians, the depth of knowledge contained in Shakespeare's plays suggests a well-traveled writer from a wealthy background with a university education, not a countryside writer like Shakespeare. But in fact, there is not much substance to such speculation, and most anti-Stratfordian arguments can be refuted with a little background about Shakespeare's time and upbringing.

First of all, those who doubt Shakespeare's authorship often point to his common birth and brief education as stumbling blocks to his writerly genius. Although it is true that Shakespeare did not come from a noble class, his father was a very *successful* glove-maker and his mother was from a very wealthy land-owning family—so while Shakespeare may have had a country upbringing, he was certainly from a well-off family and would have been educated accordingly. Also, even though he did not attend university, grade school education in Shakespeare's time was actually quite rigorous and exposed students to classic drama through writers like Seneca and Ovid. It is not unreasonable to believe that Shakespeare received a very solid foundation in poetry and literature from his early schooling.

Next, anti-Stratfordians tend to question how Shakespeare could write so extensively about countries and cultures he had never visited before (for instance, several of his most famous works like *Romeo and Juliet* and *The Merchant of Venice* were set in Italy, on the opposite side of Europe!). But again, this criticism does not hold up under scrutiny. For one thing, Shakespeare was living in London, a bustling metropolis of international trade, the most populous city in England, and a political and cultural hub of Europe. In the daily crowds of people, Shakespeare would certainly have been able to meet travelers from other countries and hear firsthand accounts of life in their home country. And, in addition to the influx of information from world travelers, this was also the age of the printing press, a jump in technology that made it possible to print and circulate books much more easily than in the past. This also allowed for a freer flow of information across different countries, allowing people to read about life and ideas from throughout Europe. One needn't travel the continent in order to learn and write about its culture.

7. What is the main purpose of this article?
   a. To explain two sides of an argument and allow readers to choose which side they agree with.
   b. To encourage readers to be skeptical about the authorship of famous poems and plays.
   c. To give historical background about an important literary figure.
   d. To criticize a theory by presenting counterevidence.

8. Which sentence contains the author's thesis?
   a. People who argue that William Shakespeare is not responsible for the plays attributed to his name are known as anti-Stratfordians.
   b. But in fact, there is not much substance to such speculation, and most anti-Stratfordian arguments can be refuted with a little background about Shakespeare's time and upbringing.
   c. It is not unreasonable to believe that Shakespeare received a very solid foundation in poetry and literature from his early schooling.
   d. Next, anti-Stratfordians tend to question how Shakespeare could write so extensively about countries and cultures he had never visited before.

9. How does the author respond to the claim that Shakespeare was not well-educated because he did not attend university?
   a. By insisting upon Shakespeare's natural genius
   b. By explaining grade school curriculum in Shakespeare's time
   c. By comparing Shakespeare with other uneducated writers of his time
   d. By pointing out that Shakespeare's wealthy parents probably paid for private tutors

10. What does the word "bustling" in the third paragraph most nearly mean?
    a. Busy
    b. Foreign
    c. Expensive
    d. Undeveloped

11. What can be inferred from the article?
    a. Shakespeare's peers were jealous of his success and wanted to attack his reputation.
    b. Until recently, classic drama was only taught in universities.
    c. International travel was extremely rare in Shakespeare's time.
    d. In Shakespeare's time, glove-makers were not part of the upper class.

12. Why does the author mention *Romeo and Juliet*?
    a. It is Shakespeare's most famous play.
    b. It was inspired by Shakespeare's trip to Italy.
    c. It is an example of a play set outside of England.
    d. It was unpopular when Shakespeare first wrote it.

13. Which statement would the author probably agree with?
    a. It is possible to learn things from reading rather than from firsthand experience.
    b. If you want to be truly cultured, you need to travel the world.
    c. People never become successful without a university education.
    d. All of the world's great art comes from Italy.

Questions 14-17 are based on the following passage which is adapted from Abraham Lincoln's Address Delivered at the Dedication of the Cemetery at Gettysburg, November 19, 1863:

Four score and seven years ago our fathers brought forth on this continent, a new nation, conceived in liberty, and dedicated to the proposition that all men are created equal.

Now we are engaged in a great civil war, testing whether that nation, or any nation so conceived and so dedicated, can long endure. We are met on a great battlefield of that war. We have come to dedicate a portion of that field, as a final resting place for those who here gave their lives that this nation might live. It is altogether fitting and proper that we should do this.

But, in a larger sense, we cannot dedicate --- we cannot consecrate that we cannot hallow --- this ground. The brave men, living and dead, who struggled here, have consecrated it, far above our poor power to add or detract. The world will little note, nor long remember what we say here, but it can never forget what they did here. It is for us the living, rather, to be dedicated here to the unfinished work which they who fought here have thus far so nobly advanced. It is rather for us to be here and dedicated to the great task remaining before us--- that from these honored dead we take increased devotion to that cause for which they gave the last full measure of devotion --- that we here highly resolve that these dead shall not have died in vain --- that these this nation, under God, shall have a new birth of freedom--- and that government of people, by the people, for the people, shall not perish from the earth.

14. The best description for the phrase "Four score and seven years ago" is?
    a. A unit of measurement
    b. A period of time
    c. A literary movement
    d. A statement of political reform

15. Which war is Abraham Lincoln referring to in the following passage? "Now we are engaged in a great civil war, testing whether that nation, or any nation so conceived and so dedicated, can long endure."
    a. World War I
    b. The War of Spanish Succession
    c. World War II
    d. The American Civil War

16. What message is the author trying to convey through this address?
    a. The audience should perpetuate the ideals of freedom that the soldiers died fighting for.
    b. The audience should honor the dead by establishing an annual memorial service.
    c. The audience should form a militia that would overturn the current political structure.
    d. The audience should forget the lives that were lost and discredit the soldiers.

17. What is the effect of Lincoln's statement in the following passage? "But, in a larger sense, we cannot dedicate --- we cannot consecrate that we cannot hallow --- this ground. The brave men, living and dead, who struggled here, have consecrated it, far above our poor power to add or detract."
    a. His comparison emphasizes the great sacrifice of the soldiers who fought in the war.
    b. His comparison serves as a remainder of the inadequacies of his audience.
    c. His comparison serves as a catalyst for guilt and shame among audience members.
    d. His comparison attempts to illuminate the great differences between soldiers and civilians.

Questions 18-20 are based upon the following passage:

The Global Water Crisis

For decades, the world's water supply has been decreasing. At least ten percent of the world population, or over 780 million people, do not have access to potable water. They have to walk for miles, carrying heavy buckets in intense heat in order to obtain the essential life source that comes freely from our faucets.

We are in a global water crisis. Only 2.5% of the water on Earth is suitable for drinking, and over seventy percent of this water is frozen in the polar ice caps, while much of the rest is located deep underground. This leaves a very small percentage available for drinking, yet we see millions of gallons of water wasted on watering huge lawns in deserts like Arizona, or on running dishwashers that are only half-full, or on filling all the personal pools in Los Angeles, meanwhile people in Africa are dying of thirst.

In order to reduce water waste, Americans and citizens of other first world countries should adhere to the following guidelines: run the dishwasher only when it is full, do only full loads of laundry, wash the car with a bucket and not with a hose, take showers only when necessary, swim in public pools, and just be <u>cognizant </u>of how much water they are using in general. Our planet is getting thirstier by the year, and if we do not solve this problem, our species will surely perish.

18. Which of the following best supports the assertion that we need to limit our water usage?
    a. People are wasting water on superfluous things.
    b. There is very little water on Earth suitable for drinking.
    c. At least ten percent of the world population does not have access to drinking water.
    d. There is plenty of drinking water in first world countries, but not anywhere else.

19. Which of the following, if true, would challenge the assertion that we are in a global water crisis?
    a. There are abundant water stores on Earth that scientists are not reporting.
    b. Much of the water we drink comes from rain.
    c. People in Africa only have to walk less than a mile to get water.
    d. Most Americans only run the dishwasher when it is full.

20. Which of the following is implicitly stated within the following sentence? "This leaves a very small percentage available for drinking, yet we see millions of gallons of water wasted on watering huge lawns in deserts like Arizona, or on running dishwashers that are only half-full, or on filling all the personal pools in Los Angeles, meanwhile people in Africa are dying of thirst."

a. People run dishwashers that are not full.
b. People in Africa are dying of thirst.
c. People take water for granted.
d. People should stop watering their lawns.

Writing

Questions 1–9 are based on the following passage:

(1) The name "Thor" has always been associated with great power. (2) Arguably, Norse Mythologies most popular and powerful god is Thor of the Aesir. (3) My first experience of Thor was not like most of today's generation. (4) I grew up reading Norse mythology where Thor wasn't a comic book superhero, but even mightier. (5) There are stories of Thor destroying mountains, defeating scores of giants and lifting up the world's largest creature the Midgard Serpent. (6) But always, Thor was a protector.

(7) Like in modern comics and movies, Thor was the god of thunder and wielded the hammer Mjolnir however there are several differences between the ancient legend and modern hero. (8) For example, Loki, the god of mischief, isn't Thor's brother. (9) Loki is actually Thor's servant, but this doesn't stop the trickster from causing chaos, chaos that Thor has to then quell. (10) In all of his incarnations, Thor is a god that reestablishes order by tempering the chaos around him. (11) This is also symbolized in his prized weapon Mjolnir a magic hammer. (12) A hammer is both a weapon and a tool, but why would a god favor a seemingly everyday object?

(13) A hammer is used to shape metal and create change. (14) The hammer tempers raw iron, ore that is in an chaotic state of impurities and shapelessness, to create an item of worth. (15) Thus, a hammer is in many ways a tool that brings a kind of order to the world—like Thor. (16) Hammers were also tools of everyday people, which further endeared Thor to the common man. (17) Therefore, it's no surprise that Thor remains an iconic hero to this day.

(18) I began thinking to myself, why is Thor so prominent in our culture today even though many people don't follow the old religion? (19) Well the truth is that every culture throughout time, including ours, needs heroes. (20) People need figures in their lives that give them hope and make them aspire to be great. (21) We need the peace of mind that chaos will eventually be brought to order and that good can conquer evil. (23) Thor was a figure of hope and remains so to this day.

1. Which of the following would be the best choice for the underlined portion of Sentence 2 (reproduced below)?

    <u>Arguably, Norse Mythologies most</u> popular and powerful god is Thor of the Aesir.

    a. Leave it as is
    b. Arguably Norse Mythologies most
    c. Arguably, Norse mythology's most
    d. Arguably, Norse Mythology's most

2. Which of the following would be the best choice for the underlined portion of Sentence 4 (reproduced below)?

    I grew up reading Norse mythology where <u>Thor wasn't a comic book superhero, but even mightier.</u>

    a. Leave it as is
    b. Thor wasn't a comic book superhero. He was even mightier.
    c. Thor wasn't a comic book superhero but even mightier.
    d. Thor wasn't a comic book superhero, he was even mightier.

3. Which of the following would be the best choice for the underlined portion of Sentence 5 (reproduced below)?

    There are stories of Thor destroying mountains, <u>defeating scores of giants and lifting up the world's largest creature the Midgard Serpent.</u>

    a. Leave it as is
    b. defeating scores of giants, and lifting up the world's largest creature, the Midgard Serpent.
    c. defeating scores of giants, and lifting up the world's largest creature the Midgard Serpent.
    d. defeating scores, of giants, and lifting up the world's largest creature the Midgard Serpent.

4. Which of the following would be the best choice for the underlined portion of Sentence 7 (reproduced below)?

    Like in modern comics and movies, Thor was the god of thunder and wielded <u>the hammer Mjolnir however there are several differences</u> between the ancient legend and modern hero.

    a. Leave it as is
    b. the hammer Mjolnir, however there are several differences
    c. the hammer Mjolnir. However there are several differences
    d. the hammer Mjolnir. However, there are several differences

5. Which of the following would be the best choice for Sentence 8 (reproduced below)?

    For example, Loki, the god of mischief, isn't Thor's brother.

    a. Leave it as is
    b. For example, Loki the god of mischief isn't Thor's brother.
    c. For example, Loki the god of mischief, isn't Thor's brother.
    d. For example Loki, the god of mischief, isn't Thor's brother.

6. Which of the following would be the best choice for Sentence 11 (reproduced below)?

This is also symbolized in his prized weapon Mjolnir a magic hammer.

a. Leave it as is
b. This is also symbolized in his prized weapon, Mjolnir a magic hammer.
c. This is also symbolized in his prized weapon, Mjolnir, a magic hammer.
d. This is also symbolized in his prized weapon Mjolnir, a magic hammer.

7. Which of the following would be the best choice for the underlined portion of Sentence 12 (reproduced below)?

A hammer is both a weapon and a <u>tool, but why would a god favor a seemingly everyday object?</u>

a. Leave it as is
b. tool; why would a god favor a seemingly everyday object?
c. tool, but, why would a god favor a seemingly everyday object?
d. tool, however, why would a god favor a seemingly everyday object?

8. Which of the following would be the best choice for the underlined portion of Sentence 14 (reproduced below)?

The hammer tempers raw iron, <u>ore that is in an chaotic state of impurities and shapelessness,</u> to create an item of worth.

a. Leave it as is
b. ore that is in a chaotic state of impurities and shapelessness
c. ore that has the impurities and shapelessness of a chaotic state
d. ore that is in an chaotic state, of impurities and shapelessness,

9. Which of the following would be the best choice for the underlined portion of Sentence 19 (reproduced below)?

<u>Well the truth is that every culture throughout time, including ours,</u> needs heroes.

a. Leave it as is
b. Well, the truth is, every culture throughout time, including ours,
c. Well, every culture throughout time, including ours, in truth
d. Well, the truth is that every culture throughout time, including ours,

Questions 10–36 are based on the following passage:

(1) Quantum mechanics, which describes how the universe works on its smallest scale, is inherently weird. (2) Even the founders of the field including Max Planck, Werner Heisenberg, and Wolfgang Pauli unsettled by the new theory's implications. (3) Instead of a deterministic world where everything can be predicted by equations, events at the quantum scale are purely probabilistic. (4) Every outcome exist simultaneously, while the actual act of observation forces nature to choose one path.

(5) In our everyday lives, this concept of determinism, is actually expressed in the thought experiment of Schrödinger's cat. (6) Devised by Erwin Schrödinger, one of the founders of quantum mechanics, it's purpose is to show how truly strange the framework is. (7) Picture a box containing a cat, a radioactive element, and a vial of poison. (8) If the radioactive element decays, it will release the poison and kill the cat. (9) The box is closed, so there is no way for anyone outside to know what is happening inside. (10) Since the cat's status—alive and dead— are mutually exclusive, only one state can exist. (11) What quantum mechanics says however is that the cat is simultaneously alive and dead, existing in both states until the box's lid is removed and one outcome is chosen.

(12) Further confounding our sense of reality, Louis de Broglie proposed that, on the smallest scales, particles and waves are indistinguishable. (13) This builds on Albert Einstein's famous theory that matter and energy are interchangeable. (14) Although there isn't apparent evidence for this in our daily lives, various experiments have shown the validity of quantum mechanics. (15) One of the most famous experiments is the double-slit experiment, which initially proved the wave nature of light. (16) When shone through parallel slits onto a screen, light creates a interference pattern of alternating bands of light and dark. (17) But when electrons were fired at the slits, the act of observation changed the outcome. (18) If observers monitored which slit the electrons travelled through, only one band was seen on the screen. (19) This is expected, since we know electrons act as particles. (20) However, when they monitored the screen only, an interference pattern is created—implying that the electrons behaved as waves!

10. Which of the following would be the best choice for the underlined portion of Sentence 2 (reproduced below)?

Even the founders of the field including Max Planck, Werner Heisenberg, and Wolfgang Pauli unsettled by the new theory's implications.

a. Leave it as is
b. including Max Planck, Werner Heisenberg, and Wolfgang Pauli; unsettled by the new theory's implications.
c. including Max Planck, Werner Heisenberg, and Wolfgang Pauli were unsettled by the new theories' implications.
d. including Max Planck, Werner Heisenberg, and Wolfgang Pauli were unsettled by the new theory's implications.

11. Which of the following would be the best choice for the underlined portion of Sentence 3 (reproduced below)?

Instead of a deterministic world where everything can be predicted by equations, events at the quantum scale are purely probabilistic.

a. Leave it as is
b. Instead, of a deterministic world where everything can be predicted by equations,
c. Instead of a deterministic world where everything can be predicting by equations,
d. Instead of a deterministic world, where everything can be predicted by equations,

12. Which of the following would be the best choice for the underlined portion of Sentence 4 (reproduced below)?

Every outcome exist simultaneously, while the actual act of observation forces nature to choose one path.

a. Leave it as is
b. Each of these outcome exist simultaneously,
c. Every outcome, existing simultaneously,
d. Every outcome exists simultaneously,

13. Which of the following would be the best choice for the underlined portion of Sentence 5 (reproduced below)?

In our everyday lives, this concept of determinism, is actually expressed in the thought experiment of Schrödinger's cat.

a. Leave it as is
b. this concept of determinism is actually expressed
c. this, concept of determinism, is actually expressed
d. this concept of determinism, is expressed actually

14. Which of the following would be the best choice for the underlined portion of Sentence 6 (reproduced below)?

Devised by Erwin Schrödinger, one of the founders of quantum mechanics, it's purpose is to show how truly strange the framework is.

a. Leave it as is
B. its purposes is to show how truly strange
c. its purpose is to show how truly strange
d. it's purpose, showing how truly strange

15. Which of the following would be the best choice for Sentence 8 (reproduced below)?

If the radioactive element decays, it will release the poison and kill the cat.

a. Leave it as is
b. If, the radioactive element decays, it will release the poison and kill the cat.
c. If the radioactive element decays. It will release the poison and kill the cat.
d. If the radioactive element decays, releasing the poison and kill the cat.

16. Which of the following would be the best choice for the underlined portion of Sentence 11 (reproduced below)?

> <u>What quantum mechanics says however is</u> that the cat is simultaneously alive and dead, existing in both states until the box's lid is removed and one outcome is chosen.

a. Leave it as is
b. What quantum mechanics says however, is
c. What quantum mechanics says. However, is
d. What quantum mechanics says, however, is

17. Which of the following would be the best choice for the underlined portion of Sentence 12 (reproduced below)?

> <u>Further confounding our sense of reality, Louis de Broglie proposed that, on the smallest scales, particles and waves are indistinguishable.</u>

a. NO CHANGE
b. Further confounding our sense of reality Louis de Broglie proposed that on the smallest scales, particles and waves are indistinguishable.
c. Further confounding our sense of reality, Louis de Broglie proposed that on the smallest scales, particles and waves are indistinguishable.
d. Further, confounding our sense of reality, Louis de Broglie proposed that, on the smallest scales, particles and waves are indistinguishable.

18. Which of the following would be the best choice for the underlined portion of Sentence 16 (reproduced below)?

> When shone through parallel slits onto a screen, <u>light creates a interference</u> pattern of alternating bands of light and dark.

a. Leave it as is
b. light created an interference
c. lights create a interference
d. light, creating an interference,

Questions 19–25 are based on the following passage:

(1) Our modern society would actually look down on some of Plato's ideas in *The Republic.* (2) But why? (3) Certainly his ideas could help create a more orderly and fair system, but at what cost? (4) The simple truth is that in many of his examples, we see that Plato has taken the individual completely out of the equation. (5) Plato's ideal society is one that places human desire aside to focuses on what will benefit the entire community. (6) To enforce these ideas, Plato seeks to use government to regulate and mandate these rules. (7) This may seem to equalize the population, its possible that this is actually the greatest breech of freedom.

(8) Today, people would think Plato's suggestion to confiscate citizens children and place them in different homes is utterly barbaric. (9) We cannot imagine the pain of losing one's own child to be raised by others. (10) In a modern trial, the judge and jury would see these as kidnapping

and ultimately condemn the philosopher. (11) As parents and emotional beings, this seems like cruelty. (12) However, the reason Plato makes this suggestion is to benefit both society and the individual. (13) The reason for this confiscation and placement is for the child to grow to the best of their potential. (14) If the child is brilliant and placed with a similar family than the child will have access to more opportunities. (15) Prosperous or intellectual parents could help develop the child's talents.

19. Which of the following would be the best choice for the underlined portion of Sentence 1 (reproduced below)?

Our modern society <u>would actually look down on some</u> of Plato's ideas in The Republic.

a. Leave it as is
b. would actually look down upon some
c. would actually be looking down on some
d. would actually look down on something

20. Which of the following would be the best choice for Sentence 3 (reproduced below)?

Certainly his ideas could help create a more orderly and fair system, but at what cost?

a. Leave it as is
b. Certainly, his ideas could help create a more orderly and fair system, at what cost?
c. Certainly, his ideas could help create a more orderly and fair system, but at what cost?
d. Certainly his ideas could help create a more orderly and fair system, at what cost?

21. Which of the following would be the best choice for the underlined portion of Sentence 5 (reproduced below)?

Plato's ideal society is one that places human <u>desire aside to focuses on</u> what will benefit the entire community.

a. Leave it as is
b. desire aside to focus on
c. desire aside focusing on
d. desire aside, focuses on

22. Which of the following would be the best choice for the underlined portion of Sentence 7 (reproduced below)?

This may seem to <u>equalize the population, its possible that this is</u> actually the greatest breech of freedom.

a. Leave it as is
b. equalize the population it's possible that this is
c. equalize the population, but it's possible that this is
d. equalize the population, however possible that this is

23. Which of the following would be the best choice for the underlined portion of Sentence 8 (reproduced below)?

Today, people would think Plato's suggestion to confiscate citizens children and place them in different homes is utterly barbaric.

a. Leave it as is
b. suggestion for confiscating citizens children
c. suggestion for confiscating citizen's children
d. suggestion to confiscate citizens' children

24. Which of the following would be the best choice for the underlined portion of Sentence 10 (reproduced below)?

In a modern trial, the judge and jury would see these as kidnapping and ultimately condemn the philosopher.

a. Leave it as is
b. jury would see this as kidnapping and
c. jury would see these kidnappings
d. jury would see those as kidnapping and.

25. Which of the following would be the best choice for the underlined portion of Sentence 14 (reproduced below)?

If the child is brilliant and placed with a similar family than he will have access to more opportunities.

a. Leave it as is
b. similar family, than the child will have access to
c. similar family then the child will access to
d. similar family, the child will have access to

Written Essay

Bob's lawnmower shop has an excess supply of lawnmowers at the end of the summer season for the first time in years. Bob's son reasons that this excess supply should be heavily discounted and sold as quickly as possible to make room for winter equipment such as snowblowers and snowmobiles to maximize profits and revenue. Bob wants to keep the unsold lawnmowers and attempt to sell them again next year when demand returns.

Write a response in which you discuss what questions should be asked and resolved in order to validate Bob's strategy as a valid one versus his son's idea. Be sure to explain how easy those answers are to obtain, what assumptions they involve, and how the answers to the questions support Bob's strategy and confirm that it has a high probability of success.

Answer Explanations #2

Arithmetic

1. C: Dividing by 100 means shifting the decimal point of the numerator to the left by 2. The result is 6.6 and rounds to 7.

2. B: Because these decimals are all negative, the number that is the largest will be the number whose absolute value is the smallest, as that will be the negative number with the least value. Thus, it will be the "least negative" (closest to zero). To figure out which number has the smallest absolute value, look at the first non-zero digits. The first non-zero digit in Choice B is in the hundredths place. The other three all have non-zero digits in the tenths place, so Choice B is closest to zero; thus, it is the largest of the four negative numbers.

3. C: To solve for the value of b, both sides of the equation need to be equalized.

Start by cancelling out the lower value of -4 by adding 4 to both sides:

$$5b - 4 = 2b + 17$$
$$5b - 4 + 4 = 2b + 17 + 4$$
$$5b = 2b + 21$$

The variable b is the same on each side, so subtract the lower 2b from each side:

$$5b = 2b + 21$$
$$5b - 2b = 2b + 21 - 2b$$
$$3b = 21$$

Then divide both sides by 3 to get the value of b:

$$3b = 21$$

$$\frac{3b}{3} = \frac{21}{3}$$

$$b = 7$$

4. C: $\frac{1}{3}$ of the shirts sold were patterned. Therefore, $1 - \frac{1}{3} = \frac{2}{3}$ of the shirts sold were solid. Anytime "of" a quantity appears in a word problem, multiplication should be used.

Therefore:

$$192 \times \frac{2}{3} = \frac{192 \times 2}{3} = \frac{384}{3} = 128 \text{ solid shirts were sold}$$

The entire expression is $192 \times \left(1 - \frac{1}{3}\right)$.

5. C: The first step is to depict each number using decimals. $\frac{91}{100} = 0.91$

Dividing the numerator by denominator of $\frac{4}{5}$ to convert it to a decimal yields 0.80, while $\frac{2}{3}$ becomes 0.66 recurring. Rearrange each expression in ascending order, as found in answer C.

6. A: 13/5

Set up the division problem.

$$\frac{5}{13} \div \frac{25}{169}$$

Flip the second fraction and multiply.

$$\frac{5}{13} \times \frac{169}{25}$$

Simplify and reduce with cross multiplication.

$$\frac{1}{1} \times \frac{13}{5}$$

Multiply across the top and across the bottom to solve.

$$\frac{1 \times 13}{1 \times 5} = \frac{13}{5}$$

7. D: Start by taking a common denominator of 30.

$$\frac{14}{15} = \frac{28}{30}, \frac{3}{5} = \frac{18}{30}, \frac{1}{30} = \frac{1}{30}$$

Add and subtract the numerators for the next step.

$$\frac{28}{30} + \frac{18}{30} - \frac{1}{30} = \frac{28 + 18 - 1}{30} = \frac{45}{30} = \frac{3}{2}$$

In the last step the 15 is factored out from the numerator and denominator.

8. C: A dollar contains 20 nickels. Therefore, if there are 12 dollars' worth of nickels, there are $12 \times 20 = 240$ nickels. Each nickel weighs 5 grams. Therefore, the weight of the nickels is $240 \times 5 = 1,200$ grams. Adding in the weight of the empty piggy bank, the filled bank weighs 2,250 grams.

9. D: 3 must be multiplied times $27\frac{3}{4}$. In order to easily do this, the mixed number should be converted into an improper fraction.

$$27\frac{3}{4} = \frac{27 * 4 + 3}{4} = \frac{111}{4}$$

Therefore, Denver had approximately $\frac{3 x 111}{4} = \frac{333}{4}$ inches of snow. The improper fraction can be converted back into a mixed number through division.

$$\frac{333}{4} = 83\frac{1}{4} \text{ inches}$$

10. C: The first step in solving this problem is expressing the result in fraction form. Separate this problem first by solving the division operation of the last two fractions. When dividing one fraction by another, invert or flip the second fraction and then multiply the numerator and denominator.

$$\frac{7}{10} \times \frac{2}{1} = \frac{14}{10}$$

Next, multiply the first fraction with this value:

$$\frac{3}{5} \times \frac{14}{10} = \frac{42}{50}$$

Decimals are expressions of 1 or 100%, so multiply both the numerator and denominator by 2 to get the fraction as an expression of 100.

$$\frac{42}{50} \times \frac{2}{2} = \frac{84}{100}$$

In decimal form, this would be expressed as 0.84.

11. B: Since $850 is the price *after* a 20% discount, $850 represents 80% of the original price. To determine the original price, set up a proportion with the ratio of the sale price (850) to original price (unknown) equal to the ratio of sale percentage:

$$\frac{850}{x} = \frac{80}{100}$$

12. C: We are trying to find x, the number of red cans. The equation can be set up like this:

$$x + 2(10 - x) = 16$$

The left x is actually multiplied by $1, the price per red can. Since we know Jessica bought 10 total cans, $10 - x$ is the number blue cans that she bought. We multiply the number of blue cans by $2, the price per blue can.

That should all equal $16, the total amount of money that Jessica spent. Working that out gives us:

$$x + 20 - 2x = 16$$

$$20 - x = 16$$

$$x = 4$$

13. B: 300 miles in 4 hours is 300/4 = 75 miles per hour. In 1.5 hours, the car will go 1.5 × 75 miles, or 112.5 miles.

14. C: One apple/orange pair costs $3 total. Therefore, Jan bought 90/3 = 30 total pairs, and hence, she bought 30 oranges.

15. D: First, the train's journey in the real word is 3 x 50 = 150 miles. On the map, 1 inch corresponds to 10 miles, so there is 150/10 = 15 inches on the map.

16. B: The total trip time is 1 + 3.5 + 0.5 = 5 hours. The total time driving is 1 + 0.5 = 1.5 hours. So, the fraction of time spent driving is 1.5/5 or 3/10. To get the percentage, convert this to a fraction out of 100. The numerator and denominator are multiplied by 10, with a result of 30/100. The percentage is the numerator in a fraction out of 100, so 30%.

17. A: First simplify the larger fraction by separating it into two. When dividing one fraction by another, remember to *invert* the second fraction and multiply the two as follows:

$$\frac{5}{7} \times \frac{11}{9}$$

The resulting fraction $\frac{55}{63}$ cannot be simplified further, so this is the answer to the problem.

18. A: The tip is not taxed, so he pays 5% tax only on the $10. 5% of $10 is $0.05 \times 10 = \$0.50$. Add up $10 + \$2 + \$0.50$ to get $12.50.

19. A: The first step is to divide up $150 into four equal parts. 150/4 is 37.5, so she needs to save an average of $37.50 per day.

20. A: The value went up by $165,000 − $150,000 = $15,000. Out of $150,000, this is $\frac{15,000}{150,000} = \frac{1}{10}$. Convert this to having a denominator of 100, the result is $\frac{10}{100}$ or 10%.

Quantitative Reasoning, Algebra, and Statistics

1. A: To solve for x the steps are as follows:

$$4x - 12$$

$$-2x, 6x - 12 = 0$$

$$6x = 12$$

$$x = 2$$

2. D: Dividing rational expressions follows the same rule as dividing fractions. The division is changed to multiplication by the reciprocal of the second fraction. This turns the expression into:

$$\frac{5x^3}{3x^2y} \times \frac{3y^9}{25}$$

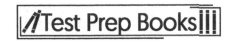

Multiplying across and simplifying, the final expression is:

$$\frac{xy^8}{5}$$

3. C: 216cm. Because area is a two-dimensional measurement, the dimensions are multiplied by a scale that is squared to determine the scale of the corresponding areas. The dimensions of the rectangle are multiplied by a scale of 3. Therefore, the area is multiplied by a scale of 3^2 (which is equal to 9):

$$24 \ cm \times 9 = 216 \ cm$$

4. B: This can be determined by finding the length and width of the shaded region. The length can be found using the length of the top rectangle, which is 18 inches, then subtracting the extra length of 4 inches and 1 inch. This means the length of the shaded region is 13 inches. Next, the width can be determined using the 6 inch measurement and subtracting the 2 inch measurement. This means that the width is 4 inches. Thus, the area is:

$$13 \ \times \ 4 = 52 \ sq. \ in.$$

5. D: When an ordered pair is reflected over an axis, the sign of one of the coordinates must change. When it's reflected over the x-axis, the sign of the x coordinate must change. The y value remains the same. Therefore, the new ordered pair is $(-3, 4)$.

6. D: 3 times the sum of a number and 7 is greater than or equal to 32 can be translated into equation form utilizing mathematical operators and numbers.

7. C: To find the mean, or average, of a set of values, add the values together and then divide by the total number of values. Each day of the week has an adult ticket amount sold that must be added together. The equation is as follows:

$$\frac{22 + 16 + 24 + 19 + 29}{5} = 22$$

8. D: First, convert the distance that Courtney already drove to feet. Because there are three feet per yard, her distance traveled thus far in yards must be multiplied by 3:

$$1,236 \times 3 = 3,708 \text{ feet}$$

If the total distance to travel is 6,292 feet, there is $6292 - 3708 = 2,584$ feet left to travel.

9. A: The probability of choosing two customers simultaneously is the same as choosing one and then choosing a second without putting the first back into the pool of customers. This means that the probability of choosing a customer who bought cherry is $\frac{35}{100}$. Then without placing them back in the pool, it would be $\frac{34}{99}$.

So, the probability of choosing 2 customers simultaneously that both bought cherry would be:

$$\frac{35}{100} \times \frac{34}{99}$$

$$\frac{1,190}{9,900}$$

$$\frac{119}{990}$$

10. B: The number line shows:

$$x > -\frac{3}{4}$$

Each inequality must be solved for x to determine if it matches the number line. Choice A of $4x + 5 < 8$ results in $x < -\frac{3}{4}$, which is incorrect. Choice C of $-4x + 5 > 8$ yields $x < -\frac{3}{4}$, which is also incorrect. Choice D of $4x - 5 > 8$ results in $x > \frac{13}{4}$, which is not correct. Choice B, $-4x + 5 < 8$ is the only choice that results in the correct answer of:

$$x > -\frac{3}{4}$$

11. B: The perimeter of a rectangle is the sum of all four sides. Therefore, the answer is:

$$P = 14 + 8\frac{1}{2} + 14 + 8\frac{1}{2}$$

$$14 + 14 + 8 + \frac{1}{2} + 8 + \frac{1}{2} = 45 \text{ square inches.}$$

12. C: The equation used to find the slope of a line when given two points is as follows:

$$slope = \frac{y_2 - y_1}{x_2 - x_1}$$

Substituting the points into the equation yields:

$$\frac{8 - (-4)}{-5 - 10}$$

$$\frac{12}{-15}$$

$$-\frac{4}{5}$$

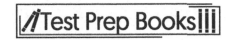

13. D: The length of LM can be found by a series of calculations:

$$KL + LM = 16 \qquad KL = 16 - LC$$

$$LM + MN = 20 \qquad MN = 20 - LM$$

$$KN = KL + MN + LM = 30$$

$$16 - LM + 20 - LM + LM = 30$$

$$36 - 30 = LM$$

$$6 = LM$$

14. B: The three angles lie on a straight line; therefore, the sum of all the angles must equal 180°. The values for angle x and angle y should be added together and subtracted from 180° to find the value for angle z as follows:

$$180 - \left(48° + 2(48°)\right) = 36°$$

15. B: Because the 65-degree angle and angle *b* sum to 180 degrees, the measurement of angle *b* is 115 degrees. Because of corresponding angles, angle *b* is equal to angle *f*. Therefore, angle *f* measures 115 degrees.

16. B: This triangle can be labeled as a right triangle because it has a right-angle measure in the corner. The Pythagorean Theorem can be used here to find the missing side lengths. The Pythagorean Theorem states that $a^2 + b^2 = c^2$, where a and b are side lengths and c is the hypotenuse. The hypotenuse, c, is equal to 35, and 1 side, a, is equal to 21. Plugging these values into the equation forms:

$$21^2 + b^2 = 35^2$$

Squaring both given numbers and subtracting them yields the equation:

$$b^2 = 784$$

Taking the square root of 784 gives a value of 28 for b. In the equation, b is the same as the missing side length x.

17. B: This problem can be solved using the Pythagorean Theorem. The triangle has a hypotenuse of 15 and one leg of 12. These values can be substituted into the Pythagorean formula to yield:

$$12^2 + b^2 = 15^2$$

$$144 + b^2 = 225$$

$$81 = b^2$$

$$b = 9$$

In this problem, b is represented by x so $x = 9$ is the correct answer.

18. C: Each value can be calculated so that they can be compared to find which one is the greatest. The mean is equal to:

$$\frac{26 + 27 + 27 + 29 + 30 + 32 + 33 + 33 + 33 + 35}{10} = 30.5$$

The median is equal to:

$$\frac{30 + 32}{2} = 31$$

The mode is equal to 33 because that number occurs 3 times in the data set. The range is equal to:

$$35 - 26 = 9$$

Therefore, the mode is the greatest value of the answer choices.

19. B: When the number of data points provided is an even number, then the average of the two middle points is the median. Each set of responses provided should be ordered from least to greatest, and then the middle two values should be averaged together to see which set provides a median of 14. Choice *B*, when ordered, is 11, 12, 13, 15, 16, and 16. The middle two values averaged together is $\frac{13+15}{2} = 14$ miles, which is the correct answer.

20. C: These two events are mutually exclusive because the students only picked one flavor of ice cream as the favorite so a student can't have chosen two flavors. Therefore, the probability of choosing a student who likes each flavor of interest (vanilla and strawberry) should be added together. 30% of students chose vanilla, and 20% of the students chose strawberry. Expressed as percentages the probabilities are $\frac{3}{10}$ and $\frac{2}{10}$ which can be added together to find:

$$\frac{3}{10} + \frac{2}{10} = \frac{5}{10} = \frac{1}{2}$$

Advanced Algebra and Functions

1. D: The values for x and y should be plugged into the equation to find the correct answer.

$$3.2(2.6)(5.3) - 4.1(5.3) = 22.366$$

2. B: First, subtract 9 from both sides to isolate the radical. Then, cube each side of the equation to obtain:

$$2x + 11 = 27$$

Subtract 11 from both sides, and then divide by 2. The result is $x = 8$. Plug 8 back into the original equation to obtain the true statement to check the answer:

$$\sqrt[3]{16 + 11} + 9$$

$$\sqrt[3]{27} + 9$$

$$3 + 9 = 12$$

3. B: The system can be solved using substitution. Solve the second equation for *y*, resulting in:

$$y = 1 - 2x$$

Plugging this into the first equation results in the quadratic equation:

$$x^2 - 2x + 1 = 4$$

In standard form, this equation is equivalent to $x^2 - 2x - 3 = 0$ and in factored form is:

$$(x - 3)(x + 1) = 0$$

Its solutions are $x = 3$ and $x = -1$. Plugging these values into the second equation results in $y = -5$ and $y = 3$, respectively. Therefore, the solutions are the ordered pairs $(-1, 3)$ and $(3, -5)$.

4. B: The function presented is being evaluated for $x + 1$; therefore, $x + 1$ must be substituted into the original function as follows:

$$f(x + 1) = (x + 1)^2 - 3(x + 1) + 17$$

The squared portion of the function becomes $x^2 + 2x + 1$, and distributing the -3 results in:

$$f(x + 1) = x^2 + 2x + 1 - 3x - 3 + 17$$

Combining like terms results in:

$$x^2 - x + 15$$

5. C: Finding the product means distributing one polynomial to the other so that each term in the first is multiplied by each term in the second. Then, like terms can be collected. Multiplying the factors yields the expression:

$$x^3 + 5x^2 - 6x + 2x^2 + 10x - 12$$

Collecting like terms means adding the x^2 terms and adding the x terms. The final answer after simplifying the expression is:

$$x^3 + 7x^2 + 4x - 12$$

6. A: Finding the roots means finding the values of *x* when *y* is zero. The quadratic formula could be used, but in this case, it is possible to factor by hand, since the numbers -1 and 2 add to 1 and multiply to -2. So, factor $x^2 + x - 2 = (x - 1)(x + 2) = 0$, then set each factor equal to zero. Solving for each value gives the values *x* = 1 and *x* = -2.

7. C: To find the *y*-intercept, substitute zero for *x*, which gives us:

$$y = 0^{\frac{5}{3}} + (0 - 3)(0 + 1)$$

$$0 + (-3)(1) = -3$$

8. A: Simplify this to:

$$(4x^2y^4)^{\frac{3}{2}} = 4^{\frac{3}{2}}(x^2)^{\frac{3}{2}}(y^4)^{\frac{3}{2}}$$

$$4^{\frac{3}{2}} = (\sqrt{4})^3 = 2^3 = 8$$

For the other, recall that the exponents must be multiplied; this yields:

$$8x^{2 \cdot \frac{3}{2}}y^{4 \cdot \frac{3}{2}} = 8x^3y^6$$

9. A: Parallel lines have the same slope. The slope of C can be seen to be 1/3 by dividing both sides by 3. The others are in standard form $Ax + By = C$, for which the slope is given by $\frac{-A}{B}$. The slope of A is 3, the slope of B is 4. The slope of D is 1.

10. D: For manufacturing costs, there is a linear relationship between the cost to the company and the number produced, with a y-intercept given by the base cost of acquiring the means of production, and a slope given by the cost to produce one unit. In this case, that base cost is $50,000, while the cost per unit is $40. So:

$$y = 40x + 50,000$$

11. A: First solve for x, y, and z. So:

$$3x = 24$$

$$x = 8$$

$$6y = 24$$

$$y = 4$$

$$-2z = 24$$

$$z = -12$$

This means the equation would be $4(8)(4) + (-12)$, which equals 116.

12. A: The solid dot is located between -2 and -3, and the open dot is located between 1 and 2. Therefore, x is between -2.5 and 1.5, which can be converted to $-\frac{5}{2}$ and $\frac{3}{2}$. The solid dot indicates greater than or equal to, and the open dot indicates less than so the inequality is:

$$-\frac{5}{2} \leq x < \frac{3}{2}$$

13. C: Nothing is added to x and y since the center is 0 and 5^2 is 25. Choice A is not the correct answer because you do not subtract the radius from x and y. Choice B is not the correct answer because you must square the radius on the right side of the equation. Choice D is not the correct answer because you do not add the radius to x and y in the equation.

14. B: For an ordered pair to be a solution to a system of inequalities, it must make a true statement for BOTH inequalities when substituting its values for x and y. Substituting (-3,-2) into the inequalities

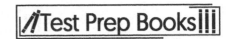

produces $(-2) > 2(-3) - 3$, which is $-2 > -9$, and $(-2) < -4(-3) + 8$, or $-2 < 20$. Both are true statements.

15. D: The shape of the scatterplot is a parabola (U-shaped). This eliminates Choices *A* (a linear equation that produces a straight line) and *C* (an exponential equation that produces a smooth curve upward or downward). The value of a for a quadratic function in standard form ($y = ax^2 + bx + c$) indicates whether the parabola opens up (U-shaped) or opens down (upside-down U). A negative value for a produces a parabola that opens down; therefore, Choice *B* can also be eliminated.

16. D: SOHCAHTOA is used to find the missing side length. Because the angle and adjacent side are known, $\tan 60 = \frac{x}{13}$. Making sure to evaluate tangent with an argument in degrees, this equation gives:

$$x = 13 \tan 60 = 13 \cdot 1.73 = 22.49$$

17. D: If $n = 2^2, n = 4$, and $m = 4^2 = 16$. This means that $m^n = 16^4$. This is the same as 2^{16}.

18. D: Each of the choices given is a linear function in the form of $y = mx + b$, where m is the slope, and b is the y-intercept. For the first function $g(x) = x - 4$, the slope is 1, and the y-intercept is -4. Placing this line on the graph would show that it does not intersect the absolute value function above. It crosses the y-axis at -4 with a slope equal to the right side of the absolute value graph, so they run parallel to one another. The second function will intersect the graph at 1 point because the y-intercept is 0. The line will run parallel to the left side of the absolute value function, with a slope of -1. For the third function, the slope is 1, and the y-intercept is 2. It will cross the absolute value graph at 1 point because it runs parallel to the right side of the graph and runs through the y-axis at 2. The fourth function has a slope of ½ and a y-intercept of 0. This line will intersect the absolute value graph at exactly 2 points. One point is at (2, 1), and the other point will be between x-values of -4 and -5. This line is shown on the graph below.

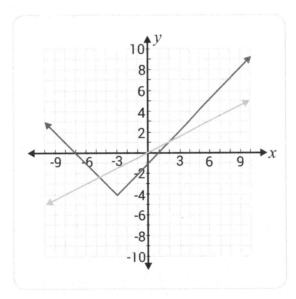

19. C: The value of $5h - 3$ cannot be a multiple of 5 because of the minus 3 at the end of the expression. Multiplying h by 5 yields a multiple of 5, but subtracting 3 means the final answer will no longer be a multiple of 5. For $10h + 20$, both the number multiplied by h and the number added to h are divisible by 5. The result will always be divisible by 5. For the final expression $h^3 + 6$, the value of h cubed yields an answer divisible by 5, but the addition of 6 means the answer is no longer divisible by 5.

20. D: The first step is to find the equation of the line that is perpendicular to $y = 2x - 3$ and passes through the point $(0, 5)$. The slope of a perpendicular line is found by the negative reciprocal of 2, which is $-\frac{1}{2}$. The y-intercept is the value of y when $x = 0$, so the y-intercept is 5. The new equation is $y = -\frac{1}{2}x + 5$. In order to find which points lie on the new line, the values of x and y can be substituted into the equation to determine if they form a true statement. For the point in A, the equation $10 = -\frac{1}{2}(-6) + 5$ makes the statement $10 = 8$. This is not a true statement, so the point $(-6, 10)$ does not lie on the line. For B, the equation $7 = -\frac{1}{2}(-2) + 5$ makes the statement $7 = 6$, which is not a true statement. Therefore, Choice B is not a point that lies on the line. For Choice C, the equation $6 = -\frac{1}{2}(2) + 5$ is the same as $6 = 4$, so it is similarly false. For D, the equation $3 = -\frac{1}{2}(4) + 5$ makes a true statement, so the point $(4, 3)$ lies on the line.

Reading Comprehension

1. D: To define and describe instances of spinoff technology. This is an example of a purpose question—*why* did the author write this? The article contains facts, definitions, and other objective information without telling a story or arguing an opinion. In this case, the purpose of the article is to inform the reader. The only answer choice that is related to giving information is answer Choice D: to define and describe.

2. A: A general definition followed by more specific examples. This organization question asks readers to analyze the structure of the essay. The topic of the essay is about spinoff technology; the first paragraph gives a general definition of the concept, while the following two paragraphs offer more detailed examples to help illustrate this idea.

3. C: They were looking for ways to add health benefits to food. This reading comprehension question can be answered based on the second paragraph—scientists were concerned about astronauts' nutrition and began researching useful nutritional supplements. Choice A in particular is not true because it reverses the order of discovery (first NASA identified algae for astronaut use, and then it was further developed for use in baby food).

4. B: Related to the brain. This vocabulary question could be answered based on the reader's prior knowledge; but even for readers who have never encountered the word "neurological" before, the passage does provide context clues. The very next sentence talks about "this algae's potential to boost brain health," which is a paraphrase of "neurological benefits." From this context, readers should be able to infer that "neurological" is related to the brain.

5. D: To give an example of valuable space equipment. This purpose question requires readers to understand the relevance of the given detail. In this case, the author mentions "costly and crucial equipment" before mentioning space suit visors, which are given as an example of something that is very valuable. A is not correct because fashion is only related to sunglasses, not to NASA equipment. B can be eliminated because it is simply not mentioned in the passage. While C seems like it could be a true statement, it is also not relevant to what is being explained by the author.

6. C: It is difficult to make money from scientific research. The article gives several examples of how businesses have been able to capitalize on NASA research, so it is unlikely that the author would agree with this statement. Evidence for the other answer choices can be found in the article: In Choice A, the author mentions that "many consumers are unaware that products they are buying are based on NASA

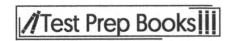

research." Choice *B* is a general definition of spinoff technology. Choice *D* is mentioned in the final paragraph.

7. D: To criticize a theory by presenting counterevidence. The author mentions anti-Stratfordian arguments in the first paragraph, but then goes on to debunk these theories with more facts about Shakespeare's life in the second and third paragraphs. *A* is not correct because, while the author does present arguments from both sides, the author is far from unbiased; in fact, the author clearly disagrees with anti-Stratfordians. *B* is also not correct because it is more closely aligned to the beliefs of anti-Stratfordians, whom the author disagrees with. *C* can be eliminated because, while it is true that the author gives historical background, the main purpose of the article is using that information to disprove a theory.

8. B: But in fact, there is not much substance to such speculation, and most anti-Stratfordian arguments can be refuted with a little background about Shakespeare's time and upbringing. The thesis is a statement that contains the author's topic and main idea. As seen in question 7, the main purpose of this article is to use historical evidence to provide counterarguments to anti-Stratfordians. *A* is simply a definition; *C* is a supporting detail, not a main idea; and *D* represents an idea of anti-Stratfordians, not the author's opinion.

9. B: By explaining grade school curriculum in Shakespeare's time. This question asks readers to refer to the organizational structure of the article and demonstrate understanding of how the author provides details to support their argument. This particular detail can be found in the second paragraph: "even though he did not attend university, grade school education in Shakespeare's time was actually quite rigorous."

10. A: Busy. This is a vocabulary question that can be answered using context clues. Other sentences in the paragraph describe London as "the most populous city in England" filled with "crowds of people," giving an image of a busy city full of people. *B* is not correct because London was in Shakespeare's home country, not a foreign one. *C* is not mentioned in the passage. *D* is not a good answer choice because the passage describes how London was a popular and important city, probably not an underdeveloped one.

11. D: In Shakespeare's time, glove-makers were not part of the upper class. Anti-Stratfordians doubt Shakespeare's ability because he was not from the upper class; his father was a glove-maker; therefore, in at least this instance, glove-makers were not included in the upper class (this is an example of inductive reasoning, or using two specific pieces of information to draw a more general conclusion).

12. C: It is an example of a play set outside of England. This detail comes from the third paragraph, where the author responds to skeptics who claim that Shakespeare wrote too much about places he never visited, so *Romeo and Juliet* is mentioned as a famous example of a play with a foreign setting. In order to answer this question, readers need to understand the author's main purpose in the third paragraph and how the author uses details to support this purpose. *A* and *D* are not mentioned in the passage, and *B* is clearly not true because the passage mentions more than once that Shakespeare never left England.

13. A: It is possible to learn things from reading rather than from firsthand experience. This inference can be made from the final paragraph, where the author refutes anti-Stratfordian skepticism by pointing out that books about life in Europe could easily circulate throughout London. From this statement, readers can conclude that the author believes it is possible that Shakespeare learned about European culture from books, rather than visiting the continent on his own. *B* is not true because the author believes that Shakespeare contributed to English literature without traveling extensively. Similarly, *C* is not a good answer because the author explains how Shakespeare got his education without university. *D* can also be eliminated because the author describes Shakespeare's genius and clearly Shakespeare is not from Italy.

14. B: A period of time. "Four score and seven years ago" is the equivalent of eighty-seven years, because the word "score" means "twenty." *A* and *C* are incorrect because the context for describing a unit of measurement or a literary movement is lacking. *D* is incorrect because although Lincoln's speech is a cornerstone in political rhetoric, the phrase "Four score and seven years ago" is better narrowed to a period of time.

15. D: Abraham Lincoln is the former president of the United States, so the correct answer is *D*, "The American Civil War." Though the U.S. was involved in World War I and II, *A* and *C* are incorrect because a civil war specifically means citizens fighting within the same country. *B* is incorrect, as "The War of Spanish Succession" involved Spain, Italy, Germany, and Holland, and not the United States.

16. A: The speech calls on the audience to consider the soldiers who died on the battlefield as ideas to perpetuate freedom so that their deaths would not be in vain. *B* is incorrect because, although they are there to "dedicate a portion of that field," there is no mention in the text of an annual memorial service. *C* is incorrect because there is no charged language in the text, only reverence for the dead. *D* is incorrect because "forget[ting] the lives that were lost" is the opposite of what Lincoln is suggesting.

17. A: Choice *A* is correct because Lincoln's intention was to memorialize the soldiers who had fallen as a result of war as well as celebrate those who had put their lives in danger for the sake of their country. Choices *B, C,* and *D* are incorrect because Lincoln's speech was supposed to foster a sense of pride among the members of the audience while connecting them to the soldiers' experiences, not to alienate or discourage them.

18. B: Choice *B* is correct because having very little drinking water on Earth is a very good reason that one should limit their water usage so that the human population does not run out of drinking water and die out. People wasting water on superfluous things does not support the fact that we need to limit our water usage. It merely states that people are wasteful. Therefore, *A* is incorrect. Answer Choice *C* may be tempting, but it is not the correct one, as this article is not about reducing water usage in order to help those who don't have easy access to water, but about the fact that the planet is running out of drinking water. Choice *D* is incorrect because nowhere in the article does it state that only first world countries have access to drinking water.

19. A: If the assertion is that the Earth does not have enough drinking water, then having abundant water stores that are not being reported would certainly challenge this assertion. Choice *B* is incorrect because even if much of the water we drink does come from rain, that means the human population would be dependent on rain in order to survive, which would more support the assertion than challenge it. Because the primary purpose of the passage is not to help those who cannot get water, then Choice *C* is not the correct answer. Even if Choice *D* were true, it does not dismiss the other ways in which people are wasteful with water, and is also not the point.

20. C: Choice *C* is correct because people who waste water on lawns in the desert, or run a half-full dishwasher, or fill their personal pools are not taking into account how much water they are using because they get an unlimited supply, therefore they are taking it for granted. Choice *A* is incorrect because it is explicitly stated within the text: "running dishwashers that are only half full." Choice *B* is also explicitly stated: "meanwhile people in Africa are dying of thirst." While Choice *D* is implicitly stated within the whole article, it is not implicitly stated within the sentence.

Writing

1. C: Choice *C* is correct, changing *Mythologies* to *mythology's*. Since one myth system is being referred to—and one particular component of it—the possessive is needed. Additionally, *Mythology's* does not need to be capitalized, since only the culture represents a proper noun. Choice *A* therefore is incorrect, with Choice *B* failing to fix the plural and Choice *D* having extraneous capitalization.

2. A: Choice *A* is correct because the sentence has no issues. While Choice *B* separates the sentence correctly, it makes more sense in this context of a direct comparison to keep the sentence intact. Choice *C* is incorrect because the sentence needs a comma after *superhero*. Choice *D* is unnecessarily long and lacks the word *but* that helps the author differentiate ideas.

3. B: Choice *B* is correct because it adds the two commas needed to clarify key subjects individually and establish a better flow to the sentence. Since *destroying mountains, defeating scores of giants*, and *lifting up the world's largest creature* are separate feats, commas are needed to separate them. Also, because *the world's largest creature* can stand alone in the sentence, a comma needs to proceed its name; *the Midgard Serpent* is not necessary to the sentence but rather provides extra information as an aside. Choice *A* is unclear and thus incorrect. Choice *C* is still missing a comma, while Choice *D* put an extraneous one in an incorrect place.

4. D: Choice *D* is correct since the sentence is lengthy as originally presented and should be split into two. Additionally, *however*, being a conjunction, needs a comma afterwards. Choice *A* is therefore incorrect due to missing punctuation. Choice *B* is an improvement but could separate the sentence's ideas better and more clearly. Choice *C* lacks the necessary comma after *However*.

5. A: Choice *A* is correct because this sentence has no issues with punctuation, content, or sentence construction. While there are three commas used, they serve to appropriately introduce an idea, an individual person, and transition into another line of thinking. Choices *B* and *C* miss commas needed to offset Loki's title as *the god of mischief*, while Choice *D* misses the comma needed to introduce the example.

6. C: Choice *C* is correct because the sentence needs two commas to emphasize the proper name of Mjolnir. Since Mjolnir is being talked about, directly addressed, and then explained, it must be flanked by commas to signify its role in the sentence. Choice *A* lacks necessary punctuation and is confusing. Choices *B* and *D* miss commas on either side of *Mjolnir*.

7. A: Choice *A* is correct, as this is an example of a compound sentence written correctly. Because of the conjunction *but* and the proceeding comma, the two independent clauses are able to form a single sentence coherently. While Choice *B* makes the question more direct, it doesn't go well with the remainder of the sentence. Choice *C* applies a comma after *but*, which is incorrect and confusing. Choice *D* inserts *however*, which is out of place and makes the sentence awkward.

8. B: Choice *B* correctly changes *an* to *a*, since *an* is only required when *a* precedes a word that begins with a vowel. Choice *A* therefore uses the incorrect form of *a*. Choice *C* fixes the issue but unnecessarily reverses the structure of the sentence, making it less direct and more confusing. Choice *D* does not fix the error and adds extraneous commas.

9. D: Choice *D* is correct, simply applying a comma after *Well* to introduce an idea. Choice *A* is therefore incorrect. Choice *B* introduces too many commas, resulting in a fractured sentence structure. Choice *C* applies a comma after *Well*, which is correct, but interrupts the flow of the sentence by switching the structure of the sentence. This makes the sentence lack fluidity and serves to confuse the reader.

10. C: Choice *C* is correct because it adds the helping verb *were* to modify *unsettled*. This allows the sentence to reflect that the founders were unsettled by the implications. Without *were* to connect *the founders* to *unsettled*, the sentence doesn't make sense. Choice *A* lacks the crucial helping verb, making it incorrect. Choice *B* is incorrect because of its unnecessary semicolon. Choice *D* changes *theories*, which is plural, to *theory's* (singular possessive), which isn't consistent with the sentence's context.

11. A: Choice *A* is correct because it contains no errors and requires no additional punctuation to form a coherent sentence. The single comma, used successfully, unites the two clauses and enables a solid grammatical structure. Choice *B* incorrectly places a comma after *Instead*, Choice *C* incorrectly changes *predicted* to *predicting*, and Choice *D* incorrectly separates *where everything can be predicted by equations* from the rest of the sentence.

12. D: Choice *D* is correct because it fixes the subject-verb disagreement with *Every outcome* and *exist*. *Exists* is third person present but also appropriate to reflect multiple outcomes, as indicated by *every outcome*. Choices *A* and *B* use *exist*, not *exists*, which makes them both incorrect. Choice *C* is fine on its own but does not fit with the rest of the sentence.

13. B: Choice *B* is correct because the comma after *determinism* isn't needed. Adding a comma in the selected area actually breaks up the independent clause of the sentence, thus compromising the overall structure of the sentence. Choices *A*, *C*, and *D* are therefore incorrect.

14. C: Choice *C* is the correct answer because it removes the contraction of *it is*, *it's*. Choice *A*, which is incorrect, originally used *it's*—note the apostrophe before *s*. *It's* simply means *it is*, while *its* (no apostrophe) shows possession. In this sentence, *its* is referring to the idea devised by Schrödinger, giving ownership of the purpose to the idea. Choice *B* is incorrect because *purpose* should remain singular. Choice *D* is incorrect because it uses *it's*.

15. A: Choice *A* is correct because the sentence is well-formed and grammatically correct. Choice *B* is incorrect because it adds an unnecessary comma after *if*. Choice *C* breaks the sentence apart, creating a sentence fragment. Choice *D* is incorrect because it changes *release* to a gerund and fails to make a coherent sentence, leaving only two dependent clauses.

16. D: Choice *D* is the correct answer. This is a tricky question, but Choice *D* is correct because, in the context of this sentence, it's important to have *however* flanked by commas. This is because the use of *however* is basically an aside to the reader, addressing an idea and then redirecting the reader to an alternative outcome or line of reasoning. Choice *A* is therefore confusing, with *however* floating in the sentence aimlessly. Choice *B* only uses one comma, which is incorrect. Choice *C* creates two incomplete sentences.

17. A: Choice *A* is correct. The sentence uses a lot of commas, but these are used effectively to highlight key points while continuing to focus on a central idea. Choice *B* is incorrect because the commas after *reality* and *that* are required. Choice *C* is incorrect because there should be a comma after *that* because *on the smallest scales* elaborates on the idea itself but not necessarily what Broglie said. Choice *D* puts a comma after *further*, which is unnecessary in this context.

18. B: Choice *B* correctly uses *an* instead of *a* to modify *interference*. The indefinite article *an* must be used before words that start with a vowel sound. The verb *created* is also in agreement with the tense

of the story. Choice *C* incorrectly changes *creates* to *create* and pluralizes *light*, which is inconsistent with the rest of the sentence. Choice *D* modifies *creates* inappropriately and adds an incorrect comma after *light*.

19. A: Choice *A* is correct because it contains no errors that mar the grammar or flow of the sentence. Choice *B* is incorrect because *upon* is incorrectly used to replace *on*. *Upon* refers specifically to a surface, which is not appropriate for this sentence. The preposition *on* is needed here. Choice *C* alters the sentence unnecessarily, confusing the tense and focus of the sentence by using *looking* instead of *look*. Choice *D* changes *some* to *something*, which makes no sense for the rest of the sentence.

20. C: Choice *C* is the correct answer because it adds a comma after *Certainly*. This is important because the author is addressing the audience before moving on to explore *his ideas*. Choice *A* is incorrect because the lack of a comma makes this sentence a run-on. Choices *B* and *D* are incorrect because, while the former applies the comma after *Certainly*, they both take away the *but* that modifies *at what costs*.

21. B: Choice *B* corrects the verb tense of *focuses*. The phrase *to focuses* is not an appropriate infinitive; the best replacement is *to focus*. The sentence is not providing the reason why *Plato's ideal society is one that places human desire*, so Choice *A* is incorrect. Using *focus* allows the prepositional phrase *to focus on* to fluidly transition into the second half of the sentence. Choice *C* is not a bad option, but it lacks the comma needed to transition into *focusing on*. Choice *D* incorrectly eliminates *to*.

22. C: Choice *C* is the correct answer because it replaces the possessive *its* with the contraction *it's* and adds a transition to an alternate viewpoint. *But* should be added as a conjunction after the comma, linking the dependent clause and independent clause. Choices *A* and *B* are therefore incorrect. Choice *D* adds a transition but incorrectly removes *its*, rendering the sentence incoherent.

23. D: Choice *D* is the correct answer because it adds an apostrophe to *citizens*, making it *citizens'*. This gives ownership to *citizens*, which clearly indicates the idea of many citizens having their children confiscated. Choices *A*, *B*, and *C* are all incorrect because they don't give ownership to citizens: the latter adds the apostrophe in a place that indicates only one citizen is impacted.

24. B: Choice *B* is correct because it changes the plural *these* to the singular *this* in order to agree with the singular noun *kidnapping*. Choice *A* is incorrect because of this number disagreement. Choice *C* would be a good alternative, but it lacks the adverb *as* to clarify the use of *kidnappings* and eliminates *and*, which serves as the connection to the rest of the sentence. Choice *D* is incorrect because *those* is plural while *kidnapping* remains singular.

25. D: Choice *D* is the correct choice because it fixes two issues with the underlined section that cause confusion for the whole sentence. First, *than* is incorrectly used in the original section. *Than* is a conjunction used to make a comparison, while *then* (acting as an adverb) serves to express a result of something, like a sequence of events. Therefore, Choices *A* and *B* are incorrect. However, therein lies the trickiness of the question because none of the answer choices use *then* without altering the meaning of the sentence (Choice *C*). Choice *D* compensates for this by replacing *than/then* with a simple comma to link the two clauses and transition from the central idea to the potential outcome of the idea (the child having more opportunities).

Dear ACCUPLACER Test Taker,

We would like to start by thanking you for purchasing this study guide for your ACCUPLACER exam. We hope that we exceeded your expectations.

Our goal in creating this study guide was to cover all of the topics that you will see on the test. We also strove to make our practice questions as similar as possible to what you will encounter on test day. With that being said, if you found something that you feel was not up to your standards, please send us an email and let us know.

We would also like to let you know about other books in our catalog that may interest you.

TSI

amazon.com/dp/162845721X

SAT

amazon.com/dp/1628457376

ACT

amazon.com/dp/1628458844

AP Biology

amazon.com/dp/1628456221

We have study guides in a wide variety of fields. If the one you are looking for isn't listed above, then try searching for it on Amazon or send us an email.

Thanks Again and Happy Testing!
Product Development Team
info@studyguideteam.com

FREE Test Taking Tips DVD Offer

To help us better serve you, we have developed a Test Taking Tips DVD that we would like to give you for FREE. **This DVD covers world-class test taking tips that you can use to be even more successful when you are taking your test.**

All that we ask is that you email us your feedback about your study guide. Please let us know what you thought about it – whether that is good, bad or indifferent.

To get your **FREE Test Taking Tips DVD**, email freedvd@studyguideteam.com with "FREE DVD" in the subject line and the following information in the body of the email:

    a. The title of your study guide.

    b. Your product rating on a scale of 1-5, with 5 being the highest rating.

    c. Your feedback about the study guide. What did you think of it?

    d. Your full name and shipping address to send your free DVD.

If you have any questions or concerns, please don't hesitate to contact us at freedvd@studyguideteam.com.

Thanks again!

Made in the USA
Middletown, DE
16 October 2020